SECOND EDITION

Nuclear Power

In Industry

A Guide for Tradesmen and Technicians

Formerly:
Atomic Energy in Industry—A Guide for Tradesmen and Technicians

LEO A. MEYER

Consultant in Vocational Education

Formerly Associate Dean for Vocational Education
Chabot College, Hayward, California

Formerly Instructor in Vocational Sheet Metal
and Apprentice Sheet Metal Classes,
Bakersfield College, Bakersfield, California

 American Technical Publishers, Inc.
Alsip, Illinois 60658

Preface to the Second Edition

This revised Second Edition of NUCLEAR POWER IN INDUSTRY has been updated to include the most current available information and illustrations. Information has been added on new techniques and measuring devices, also, information on job openings. Chapter 9, "Protection from Radiation," includes the latest OSHA standards.

NUCLEAR POWER IN INDUSTRY is written for readers who have no special training in mathematics, physics, or chemistry. Where only some knowledge of these subjects is required, concise and technically accurate summaries are provided.

Many tradesmen and technicians who will be working in the nuclear field in the near future will find their present skills greatly augmented by this informative guide to nuclear power. It will help to prepare them for the already impressive number and variety of skilled and technical jobs in the field, and for new jobs that are being created every day.

The important and interesting consequences of nuclear power must be studied, to understand the world we now live in, as well as to earn a living. This book provides a simple, accurate survey of the developments and safe utilization of nuclear power.

The Publishers wish to express their thanks to D. G. Old-field, Ph.D., biophysicist, DePaul University, for checking the manuscript.

The Publishers

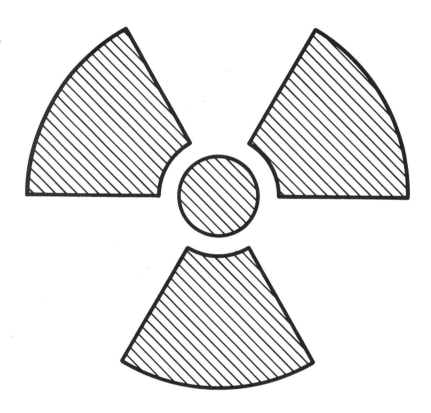

Contents

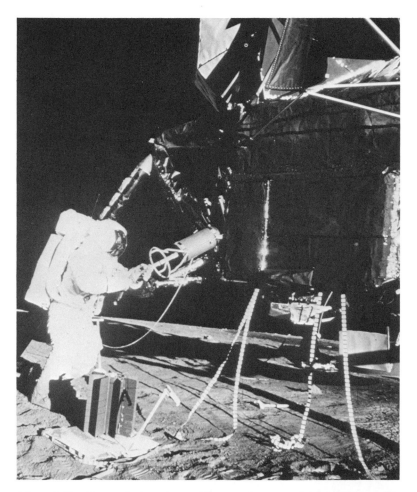

Astronaut Alan Bean unloads and assembles a nuclear generator on the moon. The generator supplies power for three years for six separate experiments. NASA.

I. Atomic Energy at Work

How is atomic energy being used to produce electric power?
How do doctors use radioactivity to treat cancer patients? How
is radioactivity being used to improve plants and animals?
How have new uses of atomic energy created new jobs and
changed old ones?

How does atomic energy—energy from the center of the atom—concern you? It concerns you because atomic energy has changed from a scientific research problem to a practical tool of industry. The first controlled release of atomic energy was achieved in 1942, but it was only after World War II that any work aimed at peaceful uses of atomic energy was done. Since then many practical ways of using atomic energy have been found. Atomic energy is a rapidly growing field that will, in the near future, be as important in our daily lives as electricity. In this chapter we will describe the sources of atomic energy, some of the many ways atomic energy is being used, and the importance of skilled workers and technicians who understand this new tool.

THREE SOURCES OF ATOMIC ENERGY

Fission

When one heavy atom is split into two light atoms, energy is released. This energy is called *atomic energy* or *nuclear energy,* and the splitting is called *nuclear fission.* Nuclear fission releases far more heat than any other energy source we now use, for the same weight of fuel. Controlled nuclear

1

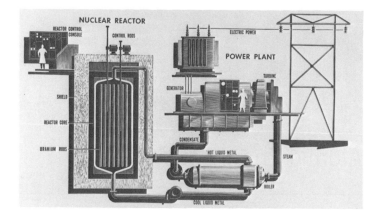

Fig. 1. Basically a reactor is an atomic furnace in which the fissioning of atoms of nuclear fuel can be put to useful work—making steam which the turbine converts to mechanical energy, which is converted by the generator to electricity. AEC.

fission is produced in a *nuclear reactor* and can be used in the same way as oil or coal as a source of heat. The heat produced by fission is used to make steam for operating turbines to generate electricity. See Figs. 1 and 2.

Radioactivity

When materials give off atomic energy in the form of penetrating rays or fast-flying particles, the materials are *radioactive,* or we can say that they contain *radioactivity*. That is, the materials contain certain atoms, called *radioisotopes,* which give off the penetrating rays. Some of these radioisotopes may

Fig. 2. The 430,000 kilowatt, San Onofre Nuclear Generating station at San Clemente, California. Southern California Edison Company.

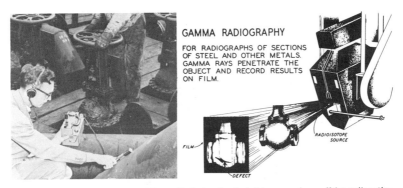

GAMMA RADIOGRAPHY

FOR RADIOGRAPHS OF SECTIONS OF STEEL AND OTHER METALS. GAMMA RAYS PENETRATE THE OBJECT AND RECORD RESULTS ON FILM.

Fig. 3. Two of the many uses industry finds for the invisible rays given off by radioactive materials. At left, a geiger counter indicates when new oil stock has arrived at a certain point in the line. California Research Corp. At right a gamma radiography setup.

have been created from the fissioning atoms themselves; others may have been created from ordinary atoms that changed because they absorbed large amounts of energy during fission. The radioisotopes give up this extra energy when they give off the penetrating rays. Fig. 3 shows two of the many ways these rays are used in industry. Almost anything that is exposed to fissioning atoms can become radioactive, but how long it remains radioactive depends on the material and how long it was exposed.

Fusion

In addition to fission and radioactivity, there is a third source of atomic energy called *fusion*. This is energy released when two light atoms are forced together to make a heavier atom. For fusion to start, extremely high temperatures (hundreds of millions of degrees) are needed. But once started, fusion creates even more heat than fission and is expected to be an important source of power in the future.

USES OF ATOMIC ENERGY

Nuclear Power

Power Plants. At the present time, nuclear power plants are a practical reality; on the other hand, building and main-

Fig. 4. The U.S. Navy's first atomic powered submarine, the Nautilus. U.S. Navy.

taining a nuclear power generator is not cheaper than building and maintaining a generator that uses conventional fuel. If the world's supply of petroleum and coal were unlimited, this application of atomic energy might not be important. But, we have found that our supply is not unlimited and that shortages are a real possibility. If we continue to use it at the rate we do now, in a few hundred years our supply of coal and petroleum will be completely gone. As conventional fuels become scarcer, the increasing cost of taking them from the earth will make them more and more expensive. Meanwhile, scientific research and the development of more efficient methods of producing nuclear energy will make nuclear power cheaper. Then atomic energy will become even more important because it is an unlimited source of power for the future.

Another big advantage of nuclear generators is that they can be used in places where conventional power plants would be impractical because of the difficulty of transporting conventional fuels. The amount of fuel needed for an atomic power plant is so small that the transportation problem would be unimportant. Prefabricated nuclear power plants small enough to be flown in by several planes and quickly assembled on the spot can be a great aid in emergency situations and also in aiding backward countries. The Army has also experimented with *portable* nuclear power plants that can be moved from place to place on a trailer. Because nuclear power plants are equally as easy to build as conventional power plants and use so little fuel, it is certain that the growth of the use of nuclear power will be tremendous. It has been estimated that by 1980 more electric power will be produced by atomic energy than is being produced now by the older methods.

Nuclear Ships. The success of the first three atomic powered submarines, the Nautilus (shown in Fig. 4), the Sea Wolf, and the Skate, has shown how valuable atomic power can be for naval vessels. Ordinary submarines are run under water by electricity. This means they have to surface frequently so that the batteries can be charged by generators run by diesel engines. Nuclear engines require no recharging, so nuclear submarines can stay submerged much longer. Also the small amount of fuel space needed leaves more room for supplies and cargo. The Navy also has nuclear aircraft carriers and cruisers now.

A major step in the development of nuclear engines was the completion of the United States Nuclear Ship Savannah, a 22,000 ton merchant ship. The Savannah, shown in Fig. 5, is partly an experimental ship, since one of the purposes of building it was to explore the problems of nuclear marine engines. But it is also highly practical; it can carry 10,000

Fig. 5. N.S. Savannah, launched at Camden, N.J., in 1959. Maritime Administration.

tons of cargo and cruise 300,000 nautical miles on one fueling.

As an example of the amount of fuel used by a nuclear powered ship, the Savannah cruised 90,000 miles on only 35 pounds of uranium-235. The Russians, Germans, Japanese and several other nations have also built nuclear-powered ships.

There are also plans being made to use nuclear propulsion for space vehicles on long flights.

Radioisotopes

Radiation produced by X-ray machines and by the naturally radioactive element, radium, have been used for years in medicine, agricultural research, and other fields. Since radioisotopes emit similar rays they can be used instead and they have several advantages. X-rays are usually produced by heavy, bulky equipment and although radium is used in small quantities, it is a rare and expensive element. Radioisotope sources can be small, like radium sources in use, but they can be manufactured in large quantities at smaller cost. In one year a nuclear reactor can produce millions of times more radioactivity through radioisotopes than all the radium ever produced.

Tracers. Radioisotopes can be produced from many substances and, even when surrounded by other material, their rays can often go through this other material, as Fig. 6 shows. The amount of radiation that comes through varies with the isotope and with the thickness and density of the material. Because of these characteristics, radioisotopes can be introduced into humans, animals, plants, or machines and their movements traced by instruments that detect radiation. Radioisotopes used in this way are called *tracers* or *tagged atoms*.

For example, fertilizer can be made radioactive and introduced into plants. Radioactive atoms in the fertilizer go through a plant in the same way as normal atoms, but the rays emitted pass through the plant and are detected by an instrument outside the plant. The instrument shows where

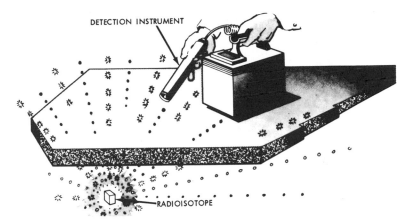

DETECTION INSTRUMENT

RADIOISOTOPE

Fig. 6. From a radioisotope source the exact location of the radioisotope and the thickness of material over it can be measured.

the fertilizer is in the plant at any time. By tracing the movement of the fertilizer through the plant it has been learned how fast fertilizer moves through a plant, the most effective way of placing fertilizer, and how the plants absorb fertilizer. Fertilizer studies are expected to result in increased crop yields. They may even lead to a speed up in plant growth to the point where more than one crop can be harvested in a season.

In the same way, atoms of food can be made radioactive and their path through human beings or animals studied. Medical men need to understand the details of normal processes in the body so that they can better treat abnormal conditions. Some of these processes are so complex that there is no way to study them except with radioactive tracers. The amount of radioactivity given off by tracers is so small that it doesn't damage the plants or animals.

Tracers have uses in mechanical systems too, as Fig. 7 shows. If tracers are introduced into the water of a pipeline under pressure, very small leaks that could not be found by ordinary pressure tests can be detected. Or small amounts of radioactive oil can be run through a large oilfield pipeline to

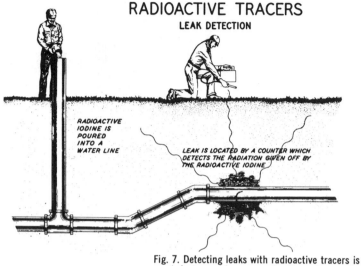

Fig. 7. Detecting leaks with radioactive tracers is much faster and less expensive than other methods. Atomic Energy Commission.

pinpoint the location of a stoppage. A man walking over the area with a detection instrument can find the exact location where the radioactive source stops moving in the pipe. This is much faster and less expensive than the usual methods of searching for a stoppage.

Treatment of Cancer. Soon after the discovery of X-rays and radium, just before 1900, doctors learned that radiation has the power to slow down or stop the growth of cancer. They have been using it for this purpose ever since. Now that radioisotopes are available, a much wider range of treatments is possible (Fig. 8). The results using isotopes have been better than they were using only X-rays and radium.

Scientists are trying now to make a radioactive substance that will be taken up only by cancer cells. If this were introduced into the body, it would go to the cancer cells and be absorbed by them but it could not be absorbed by any other cells. Then the cancer cells would receive a large dose of radiation with little danger to the other parts of the body. Eventually it might even be possible to make radioactive substances that would kill the bacteria causing certain diseases without affecting healthy tissue.

Even if not taken up only by cancer cells, often *more* of the radioactive substance is taken up by tumor cells than by healthy cells. And if this happens, the tumor can be *located* by moving a radiation detector over the body of the patient. Such devices are called *scanners*.

Improved Plants and Animals. Agriculture and animal breeding have many uses for radioisotopes. Not only are they useful as tracers for studying the processes that take place inside plants and animals but they are also being used to produce better strains. In nature most of the plants grown from seeds are like the plants that produced the seeds, but a few are radically different. These changes in the plants are called *mutations* and the plants are called *mutants*. Many mutants are worthless but some are much better than the original plants. The best ones are developed in order to obtain new plants that will produce more and better products. Cattle, sheep and similar animals are improved by the same method. The best animals are selected and used for breeding purposes.

If plants and animals are exposed to radiation the number

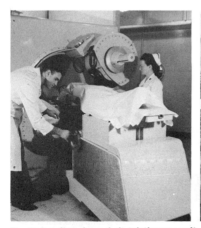 

Fig. 8. A radioactive cobalt teletherapy unit in use at Oak Ridge Associated Universities. AEC's Oak Ridge Associated Universities. | Fig. 9. Experimental greenhouse at Brookhaven National Lab. Cobalt in the center pipe irradiates plants.

of mutations can be greatly increased. Selecting and developing superior strains is a very old process but radiation makes it possible to get results much faster. In one case, for example, a project to develop rust-disease-resistant oats was successful after only about a year and a half of experimenting. This might have taken ten years or more if only natural mutations were available. Irradiated seeds are available in garden stores now, so anyone who wishes to can experiment with mutations. Fig. 9 shows an experimental greenhouse where the effects of radiation on plant growth and development are being studied.

Food Processing. It has been discovered that exposure to radiation kills the bacteria that cause food decay and makes it possible to preserve food for years without refrigeration or freezing, Fig. 10. Proper exposure to radiation does not make anything radioactive and the food is perfectly safe. Low level radiation used to extend food life is being used in experimental projects in Washington on orchard products, in Ha-

Fig. 10. Radiation extends food life. These potatoes were photographed 16 months after exposure to gamma rays. The potato in the upper left was not exposed. The others were exposed to increasing levels of radiation. The one in the lower left received the greatest exposure and is still firm. Brookhaven National Laboratory.

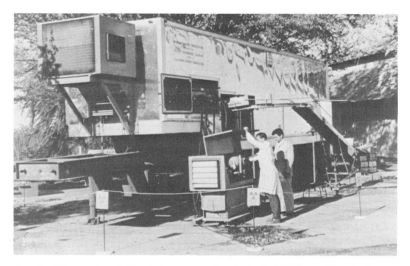

Fig. 11. This experimental irradiator was used under typical field conditions to demonstrate the ease of the process. Picked fruit is carried by conveyor through the irradiator. University of California.

waii on tropical fruit, and on board ship in the Atlantic Ocean on fish. See Fig. 11. Low level radiation merely extends the life of the food product; it does not preserve it entirely. However, this is important since in the United States, the loss on orchard products alone is near 100 million dollars each year. High level radiation will preserve food without refrigeration or freezing for many years. The greatest problem is that irradiation changes the appearance and the taste of the food. Processes are gradually being found to eliminate this problem. The Food and Drug Administration which has the responsibility of approving processed food for human consumption, has approved the radiation sterilization process for bacon, wheat, wheat flour, and potatoes.

The U. S. Army maintains a Radiation Laboratory at Natick, Massachusetts for research in food irradiation. In one year, the Army produced 30,000 pounds of radiation-sterilized bacon for its own use. In the same year 40,000 pounds of potatoes were radiation-sterilized for use on military bases.

Radioisotopes in Industry. Radioisotopes are now familiar

materials in industry. They are used for tracing materials, for wear studies, for gaging thickness, and many other uses. A comparatively new application is the use of radioisotopes as the heart of a device to detect impurities in the air. These are used as smog alarms.

Radiation Sterilization. Radiation can be used to sterilize medical supplies such as scissors, needles, bandages, sponges, etc., right through the protective wrapping or container that holds them.

Radiation Hardening. Certain plastics such as polyethylene become tougher and more resistant to chemical attack after being irradiated.

Thickness Control. Radioisotopes can be used in industry to control the thickness of a material because their rays will pass through the material in different amounts depending on the material's thickness. By placing a radioactive source under the paper, metal, or other sheet material that comes from a rolling mill and a detection instrument over the top, we can measure thickness to within a few thousandths of an inch. See Fig. 12. These measurements are very accurate and they can be made at the instant the material is rolled out, so that the mills can automatically adjust to produce the correct thickness. This method has been used in paper mills for several years with great success.

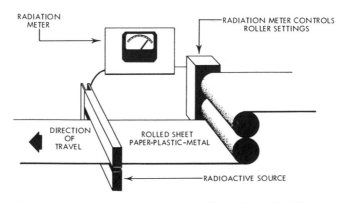

Fig. 12. Radiation thickness control setup. Atomic Energy Commission.

Radiography. If a radioisotope is placed on one side or inside of a metal casting or forging and a photographic plate on the other, more rays will be stopped by the thick sections of metal than the thin ones. The photographic plate is darkened to different degrees depending on how much radiation is coming through, so that a complete black and white shadow picture of the object, showing flaws and weak spots, is produced. Radiography is widely used now for making nondestructive tests on castings, forgings, pipes, ship and tank welds, etc.

Wear Studies. Wear studies on car engines using radioisotopes can be made in a fraction of the usual time. Parts of the engine are made radioactive; then the engine is run for a definite period of time. By draining the oil from the engine and measuring the amount of radioactive material it contains we can determine the exact amount of metal that has worn off the parts. This can be done even when the amounts are so minute that they could never be measured by ordinary methods.

Static Eliminators. In dry weather, cloth, paper, plastics and other non-conducting materials become electrically charged by friction in passing between rollers, etc. The charge can be removed by exposing the material to radiation, which produces ions in the air that will neutralize the charge.

Nuclear Batteries. By collecting the electrons given off by a radioactive substance, a simple, rugged nuclear battery can be made that keeps operating as long as the radioactive material lasts. The electrical current produced is usually very small, but the battery can operate under conditions of vacuum and high temperature that would ruin other types of batteries.

Permanent Light Sources. Certain substances give off light when irradiated; if a radioisotope is mixed with such a substance, the combination will give off a faint glow of light for as long as the radioisotope lasts. Such sources are useful for testing certain kinds of light-measuring instruments.

ATOMIC ENERGY WORK

These are just a few examples of how atomic energy is being used in agriculture, medicine and industry. As early as 1956 there were more than 32,000 locations in New York City alone where radioactive materials were in use and presently the number of these has increased tremendously. New uses for nuclear power and radioisotopes are being found every day and every discovery means more work for everyone: making, maintaining, using, and transporting atomic energy equipment and materials.

Skilled Craftsmen in Atomic Energy Work

Atomic energy has provided many new job opportunities for skilled craftsmen because of the need to fabricate special parts and equipment for experimental work, Fig. 13, the custom manufacturing of many items, the precise tolerances that must be maintained, and the great amount of complex machinery and equipment that must be serviced.

Journeymen from many skilled trades are employed in all parts of the country in installations under the supervision of the Atomic Energy Commission, and the number continues to increase. This work includes construction and maintenance of laboratories and other buildings, reactors, electrical power generating plants, construction and maintenance of electrical safety and monitoring devices, and the manufacture of reactor components and instruments. Journeymen from almost every skilled trade are employed in this work, including electricians, plumbers, pipe fitters, steam fitters, welders, machinists, carpenters, sheet metal workers, and many other occupations.

New Jobs for Technicians

Technicians of all kinds are employed in atomic energy work—as radiographers, draftsmen, computer programmers, reactor and accelerator operators, radioisotope production operators, radiation monitors, laboratory technicians, decon-

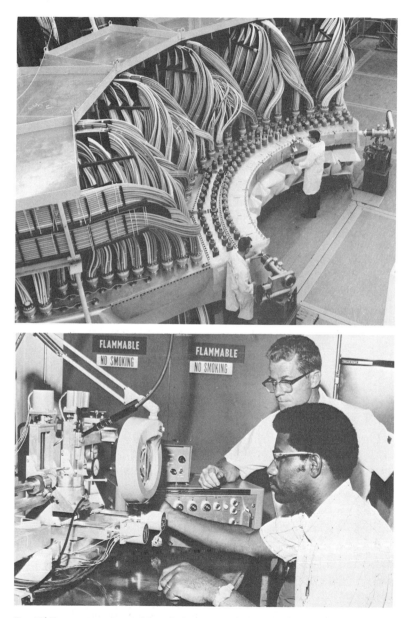

Fig. 13. **Top,** an experimental installation for research in producing electricity through fusion. Los Alamos Scientific Laboratory. **Bottom,** in atomic energy laboratories, mechanics, technicians, and engineers are needed to build, install, and test intricate experimental equipment. Savannah River Plant.

tamination men, waste treatment and waste disposal operators, and many others.

Since training requirements have not been standardized,
most employers hire workers with related skills, or high
school or junior college graduates who have had courses in
physics, chemistry or mathematics, and provide on-the-job
training. For example, photographers are hired for radiography training; mechanics for handling radioactive material
with remote control equipment. Men experienced with electrical measuring instruments and with laboratory apparatus
can learn to operate specialized radiation-producing machines
called accelerators. Decontamination and waste disposal technicians need to be familiar with electronic instruments for
radiation measurements.

Jobs Indirectly Involved with Atomic Energy

In laboratories, hospitals, plants, and other establishments
that use radioisotopes, most employees do the same work as
they would otherwise. The difference is that almost all such
work is a little more complicated because of the safety precautions that have to be taken, as you will see from the examples that follow.

Electrical Work. Electricians do the same type of work as in
other building construction and maintenance except for the
additional extra requirements for safety. Monitoring and
alarm systems, backup systems in case of power failure, and
many remote control and viewing systems require highly
skilled and expert journeyman electricians, Fig. 14.

Sheet Metal Work. Sheet metal workers will need to understand radiation because ventilation and exhaust systems play
a large part in radiation safety. Some types of radiation are
given off by dust particles that can be carried in air, so the air
in atomic energy installations has to be filtered. In laboratories, each hood is ducted to an outside exhaust and each
exhaust line has a double filtering system. At the hood a

Fig. 14. Control panels, such as this one for a high flux isotope reactor at Oak Ridge Laboratories, require installation by skilled electricians. Oak Ridge National Laboratory.

special filter is installed to filter out the radioactive particles; then a second filter is installed on the blower to filter out any remaining particles. A special monitoring system is installed above the roof line of the duct for checking to see if any radioactive materials are escaping into the exhaust air.

Plumbing and Other Trades. In plumbing also, safety precautions will make the work more complicated. In many installations special plumbing systems must be installed to prevent radioactive materials in the water from being carried down into the sewer system.

WHY STUDY ATOMIC ENERGY?

The question still remains—*"Why study atomic energy?"* Isn't it enough to know that it exists and that its use will grow? Anyone who understands something about atomic energy has a better chance of getting and keeping one of the many good jobs in this field than someone who doesn't. So many new uses of nuclear power and radioisotopes are being found that it may not be long before all jobs are related in some way to atomic energy, directly or indirectly.

Nuclear power is used to run the generator that supplies electricity for the lights and transmitter on this weather buoy located in the Gulf of Mexico. The Martin Co.

To find another answer to this question let's look at another useful tool of industry—electricity. Suppose we took a man from the most backward region of the world, who had no experience with electricity. If we put him to work on a 110 volt line, what would happen? He would most likely touch the bare wires and be "knocked on his ear"—if he were lucky enough not to be killed. What would his reaction be? After he got over being numb he would probably try to figure out what had happened. Since he knew nothing about electricity he would base his explanations on superstition and think up all sorts of wild stories to explain the jolt of electricity he got.

For a workman in any trade it is not enough to know that electricity is used extensively in industry and construction. He must know that touching the bare wires will give a shock, that standing in water while touching the bare wires could give an even more dangerous shock, and that a shock from 110 volts is not usually fatal but can sometimes be so. Above all, he must know that a shock is liable to stop a person's breathing and that artificial respiration can save a victim's life.

If, in addition to these things, he knows the basic principles of electricity, he is even better equipped to make the most of the advantages of electricity. Atomic energy is the same as electricity in this respect. A man who understands the basic

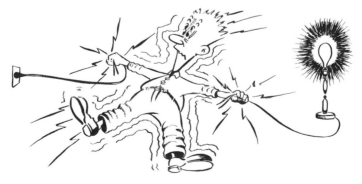

Electricity is dangerous too, if you don't know anything about it.

principles of atomic energy and the dangers involved is better equipped to do his job than one who is ignorant of them.

Another example also concerns electricity. An electrician may be highly skilled in his trade but could still run into considerable problems in trouble-shooting air conditioning and heating control systems unless he understood the different components that made up the system and the reason for each of the various controls. The old saying that "ignorance is bliss" is definitely not true here. In the case of atomic energy—as with electricity—ignorance is *danger*.

Ignorance may lead to baseless fear. You have probably heard a lot of false nonsense concerning the dangers of radioactivity such as:

If you are exposed to radiation you will become radioactive.—Wrong.

If you are exposed to radiation you cannot have children or, if you do, they will be monsters.—Wrong.

Radiation is the cause of heavy rains.— Wrong.

Radiation is the cause of droughts.— Wrong.

These stories and similar ones are originated by people who know nothing about radiation and passed on by others who know even less. None of them are true and they add to the greatest danger of radiation—*ignorance.*

Very many of us will be working around radioactive materials at some time in the future. If we remain ignorant and accept stories we hear from uninformed persons we will be in the position of a man learning about heat prostration from an Eskimo. The secret of safety in atomic energy work lies in obtaining the basic knowledge that will make us able to judge what is dangerous and what is not. We must depend on knowledge to guide us because our senses will not tell us when there is danger.

Let's look at some imaginary examples that illustrate this.

Suppose an electrician walks into a laboratory to replace a receptacle. In order to get to the receptacle he moves a small piece of metal about the size of a silver dollar and some lead blocks surrounding it. After he has installed the new outlet he carefully replaces the metal and the lead blocks in exactly the same manner as he found them. In this case, the small piece of metal was a highly active radioisotope — the lead blocks were there as a shield to absorb its rays—and the electrician has received so much radiation that it might kill him.

As a second example, suppose that a sheet metal worker enters a building on a routine job of cutting a hole in an exhaust duct and installing a new exhaust line. In the process of cutting the hole in the existing duct he cuts his hand slightly. Most of us pay no attention to a small cut. But, if the building was a laboratory handling radioactive materials, the man could receive internal radiation poisoning through the cut that would cause a long sickness or death. If he knew enough about radioactivity to report the accident to the health physicist of the laboratory, it might save his life.

As another example, let's say a plumber enters the same building to clean out a clogged trap under a sink. When he removes the trap he spills some of the water on the floor and on his shoes. He is a conscientious worker, so he takes a rag from his pocket and wipes up the mess. Then he walks out to his truck to get some new washers for the trap. He replaces the trap, cleans up, and leaves the job. But the water in this building contains radioactive waste. The plumber has tracked it through the building, making an expensive cleanup job necessary, and he is still carrying it around with him on the rag.

Think of the transportation field: suppose a truck driver transporting a metal radioisotope has a collision that breaks the isotope's shielding container. Knowing the isotope is dangerous and valuable the driver quickly picks it up and puts it in a cardboard box. He would suffer severe radiation burns on his hands from doing this and would also be exposing other people to radiation since the cardboard box would not stop the rays.

Safety around radioactive sources is always supervised by

trained health physicists. Accidents are extremely rare—the ones described above probably never happened — but they could become more frequent if employees remain ignorant of the potential dangers.

All this would make it seem that nuclear energy work is an extremely dangerous thing. Actually, it is no more dangerous than working with electricity or on a roof or scaffold. Radiation can't kill you any deader than a jolt from a 10,000 volt high tension line but everyone knows the dangers of electricity and the precautions needed. No one is afraid to work around power lines and accidents happen only through carelessness. Atomic energy is a little harder to understand than a simple electrical line, but it is no more dangerous, and the dangers that do exist can be easily understood and avoided.

QUESTIONS

See how much you have learned by completing the following statements. The answers appear at the end of the book so that you can check yourself to see which parts of the chapter you understand and which you should go over again.

1. The terms _____ energy and _____ energy are more or less interchangeable.
2. Nuclear energy is produced when a heavy atom is _____.
3. In power plants heat from nuclear energy is used to produce (a) _____ which is then used to produce (b) _____.
4. Besides being produced by atom splitting, nuclear energy is given off by _____ materials.
5. Another name for radioactive material is _____.
6. When radioactive materials are introduced into animals, plants, and machines and their movements followed by a radiation detector, they are called _____.
7. In agriculture, radioisotopes are used to produce new types of plants by increasing the rate of _____.
8. Nuclear power plants have the great advantage of requiring a very small amount of _____.

II. The Atomic Age Begins

When did scientists discover that matter, which looks solid, is actually composed of atoms so small they can't be seen, even through the most powerful microscope? How did the discovery of electricity lead to the discovery that atoms are themselves made of smaller particles? How did the discovery that matter can be turned into energy help to win World War II? If matter and energy can be turned into each other, why didn't anyone discover this before the present century? Isn't this what happens when wood is burned and turned into heat?

We are concerned with the practical part of atomic energy—what it is, how it is used, how it affects our work. You may wonder, then, why we have included a chapter on the beginnings of atomic science.

The main reason is that it is easier to understand atomic science when we see how it grew. We can see the simple and straightforward beginnings of ideas that seem a bit complicated now. We can see how the simple notions had to be changed to fit what nature is instead of what man wanted her to be. We can see how separate discoveries sometimes dovetailed with each other and slipped into place in the jigsaw puzzle of science.

THE ATOMIC THEORY

The search for an understanding of what our universe is made of led scientists to dissect matter into smaller and finer bits, and to develop theories of how these pieces exist in nature. First was the development of the atomic theory. This is

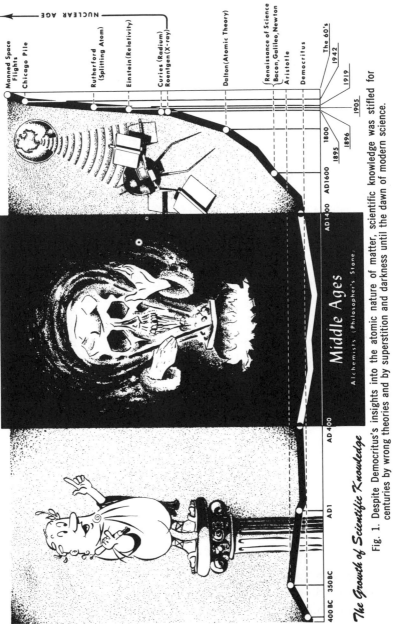

The Growth of Scientific Knowledge

Fig. 1. Despite Democritus's insights into the atomic nature of matter, scientific knowledge was stifled for centuries by wrong theories and by superstition and darkness until the dawn of modern science.

the theory that all matter is composed of *atoms,* tiny particles with empty space between them, instead of being a continuous substance. It took thousands of years to learn this, almost the whole length of recorded history. After the atomic theory was accepted, the next step was the study of atomic physics. This is the study of atoms themselves, the parts they are made of and how these parts are joined together. Nuclear physics is the study of the center of the atom, the seat of nuclear energy. Fig. 1 indicates the tremendous increase in scientific knowledge in recent years. Compared to the time it took to develop the atomic theory, nuclear physics has been with us only an instant of history.

Early Theories on the Nature of Matter

As long as we have had civilization there have been brilliant men who have tried to discover the basic construction of matter. Four hundred years before Christ was born, a Greek philosopher named Democritus worked out a theory that matter was composed of fundamental particles, which he called *atoma.* Considering that Democritus had so little evidence to go on and that he was dealing with particles too small to be seen, it is amazing that he guessed the truth.

It often happens in science, as elsewhere, that people follow wrong theories. The wrong theory, in this case, was that of Aristotle who developed the idea that matter was composed of a fundamental substance that could have four different properties, heat, cold, wetness and dryness. When matter had the properties heat and dryness, it was fire; when it had cold and dryness, it was earth; with cold and wetness it was water; and with heat and wetness, air. According to Aristotle, everything else that existed was a mixture of the four elements; fire, earth, water and air.

The Philosopher's Stone

For hundreds of years most men of science believed in Aristotle's theory and in the Middle Ages the alchemists based their search for the "philosopher's stone" on it. This was a stone that could take the four basic elements of a common

You can't get rich by turning lead into gold.

metal and rearrange them in the proper order to produce gold. The alchemists were the closest thing to scientists that existed in the Middle Ages, but they were more like modern confidence men than scientists. They spent more time tricking kings and common people out of their money than they spent in doing experiments. Still, they did do some experiments, which was a step in the right direction for making a science of the study of matter. The Greeks had answered questions about the nature of matter by pure reasoning, but they almost never tried out their notions to see if nature actually worked in the way they reasoned it should work. The alchemists tried out their notions but, because of their get-rich-quick approach and because their basic thinking on the problem was wrong, they failed.

Today we have our philosopher's stone, but it is called a *nuclear reactor*. It can change mercury into gold or gold into platinum. But gold or platinum made this way costs more than the kind found in nature. On the other hand, the alchemists never dreamed that their philosopher's stone could produce heat or light or drive giant metal fish under the ocean. The nuclear reactor does all of these things.

Modern Science

Modern science began in about the seventeenth century, when men like Bacon, Galileo, and Newton established some real scientific principles. Bacon was a philosopher who had the idea that laws of nature could only be discovered by observing nature and by setting up experiments that would show definitely which of several possible laws of nature was the correct one. Galileo discovered the laws telling how fast a falling object drops and how fast a pendulum swings. He proved these laws by performing the necessary experiments and observing the results. Newton used observations of the position of the sun and planets taken at different times to check that his law of gravitation actually could predict the motion of the planets around the sun.

Another example of the power of the new methods of observation and experimentation was the atomic theory of Dalton. This theory said more than just that all matter was made of atoms; it also gave the rules for the combination of atoms with each other. And Dalton backed up his theory with the results of experiments. The atomic theory of Dalton led later to methods for finding how much each atom weighs.

Atoms and Molecules

An *atom* is defined as a tiny building block of matter that can't be divided into smaller units without changing its basic character. Before Dalton proved his atomic theory, many people thought a piece of gold could be divided into any number of pieces, and no matter how small the pieces were, they would still be gold. But, according to the atomic theory, if you have one atom of gold and break it into two parts, the parts will not be gold. You would have one atom each of two lighter materials.

A *molecule* is a combination of two or more atoms joined together by forces between the atoms. The atoms in a molecule can be all the same or all different. One molecule of oxygen contains two atoms of oxygen. One molecule of carbon monoxide gas contains one atom of carbon and one atom of oxygen. See Fig. 2.

Elements and Compounds

Many common materials, such as iron, lead, and carbon contain atoms of only one type. These substances are called *elements*. There are 105 elements known and new ones are

Fig. 2. How to remember the relationship between atoms and molecules.

still being found. Many other materials, such as water, salt, and air, contain atoms of two or more different types. These materials are either *compounds* or *mixtures*. A compound always has one and only one definite composition in terms of the relative number of atoms of each kind present. Water is a compound of hydrogen and oxygen. It always contains two atoms of hydrogen for every atom of oxygen, never more or less. Air is not a compound but a mixture of nitrogen, oxygen, carbon dioxide, and other gases. The composition of air can certainly change — in a stuffy room, the amount of carbon dioxide goes up.

The Structure of the Atom

The suspicion that the atom is not an indivisible unit of matter, that the atom itself must contain certain smaller parts or particles, had its beginnings in the experiments of the great English physicist Faraday in 1834. His research was concerned with the passage of electricity through solutions, and showed that each molecule in solution carries a constant amount of electrical charge. But measurements on free particles, separated from their atoms, were not made until 1897. These measurements, by J. J. Thompson, showed that the particles removed from the atoms had a negative charge and that the properties of the particles were the same no matter what kind of atom they were removed from. The particles were, of course, what are now called *electrons*.

Since the electrons were negative, it was reasoned that there must be an equal amount of positive electricity in the atom to balance them. In 1911 Rutherford proposed on the basis of experimental evidence that the positive electrical particles were contained in the center or *nucleus* of the atom, and that practically all the weight of the atom was also contained in the nucleus. These heavy, positively charged particles were called *protons*. Finally Chadwick in 1932 discovered that heavy, neutral particles, which were called *neutrons,* were also contained in the nucleus of the atom.

Each atom, then, is composed of a heavy nucleus or center, which is a cluster of neutrons and protons, and a number of electrons that rotate around the nucleus, somewhat the way the earth and other planets orbit around the sun. See Fig. 3.

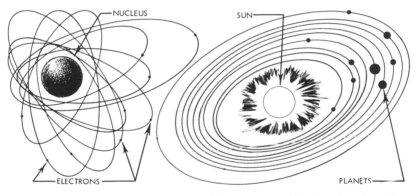

Fig. 3. Parts of the atom compared to the solar system. The atom's nucleus is like the sun; the electrons spinning around it like the planets orbiting the sun.

MASS AND ENERGY

When Albert Einstein published his theory of relativity in 1905, the world of physics was soon "shook up." In this theory he showed that while Newton's laws of motion were correct for most cases, they were no longer correct for the case when an object was moving at extremely high speeds; that is, at speeds near the speed of light (186,000 miles per second). Einstein developed a theory that was correct for very fast-moving objects and that also turned back into Newton's laws of motion again for not-so-fast-moving objects.

One tremendously important result of the theory was concerned with the relation between *mass* and *energy*. In the older physics before Einstein, mass was the amount of matter contained in an object, and energy was something the object had because it was moving or because of its position. For example, a hammer contains a certain amount of matter which is its mass. As the hammer comes down to drive in a nail, it has kinetic energy because it is moving. Before it started down, the hammer had potential energy because it was being held high. In the older physics it was thought that the mass of the hammer and how fast the hammer was moving had nothing to do with each other. Einstein showed that this idea was wrong and that if the hammer moved very fast its mass would actually get larger.

Einstein also showed that even when the hammer was not moving and not being held high, it still had a certain initial amount of energy called *rest energy* just because the mass of the hammer existed. And it followed from the theory that if some of the mass of the hammer were made to disappear, the amount of rest energy corresponding to the mass lost would have to appear as actual energy. This result is Einstein's famous law of the *equivalance of mass and energy* and is given by the simple but important formula $E = mc^2$. In this formula, E is the energy produced when a mass m disappears. The symbol c^2 means c times c, and c itself is the speed of light.

The best way to see what the formula $E = mc^2$ tells us is to put some numbers into it. The speed of light is 186,000 miles per second, which is the same as 30,000,000,000 (thirty billion) centimeters per second. So c^2 is 30,000,000,000 × 30,000,000,000 = 900,000,000,000,000,000,000. When the mass m is in grams, the energy E is in *ergs*. One erg is a fairly small amount of energy. For example, if you plucked a short hair out of your head and let it drop from a height of one centimeter (about 3/8″), the energy of that hair when it hit the ground would be about one erg. If instead of dropping the hair, we could convert all its mass (say about 1/1000 of a gram) into energy, the formula tells us that we would get 1/1000 × 900,000,000,000,000,000,000 = 900,000,000,000,-

000,000 ergs of energy. This is an enormous amount of energy—enough, for example, to blast a heavy automobile a few thousand miles into the air. The trick, of course, is to find where in nature such transformations of mass into energy occur. Ordinary chemical reactions, such as burning wood or coal or even exploding dynamite, show almost no change in mass.

What Happens When Atoms Split

The place to look for changes in mass and for large amounts of energy released is inside the atom; in fact, in the nucleus of the atom. As you learned earlier, the splitting or fission of a heavy nucleus produces energy. Also, it turns out that the masses of atoms that result from splitting are smaller than the mass of the atom that split. So from the Einstein mass-energy equivalence, we know that fission should release a huge amount of energy, and indeed it does. The same is true for fusion. When two light atoms are forced together, the mass of the product atom is less than the masses of the atoms that fused, and the mass difference is converted into energy.

What Happens When Wood Burns

When wood burns it looks as if most of the wood is turning into heat because the ashes that are left weigh much less than the wood, but this is not what happens. The fire you see is actually red-hot gas. When the gas, smoke, and ashes are all weighed together they seem to weigh exactly as much as the wood did. From experiments like this one chemists and physi-

When wood burns, the smoke, gas, and ashes produced weigh almost exactly as much as the wood did.

cists concluded that when one material changes into another, energy can change from one form to another, but no material is really destroyed in the process. In this example, energy in the form of heat is produced by the chemical changes occurring when wood turns to gas, smoke, and ashes, but the weight of material is exactly the same after burning as before.

Now it is known that when wood burns some of its mass could be destroyed and changed into energy. But this has never been measured because the difference between the weight of wood and the weight of gas, smoke and ashes is so small that it is practically undetectable.

The Discovery of X-rays and Radioactivity

A landmark in the study of the atom was passed in 1895 when Wilhelm Roentgen discovered the penetrating radiation he called *X-rays*. Roentgen was experimenting with the passage of electricity through a glass tube from which the air had been evacuated. When he turned off the lights he noticed a greenish glow from a paper screen that had been coated with barium cyanide crystals. He found that this glow was caused by rays coming from the glass tube and hitting the paper screen. The rays came from that part of the tube being struck by the electrons passing through the tube.

The property of X-rays that makes them useful is that they pass through some materials more easily than through others. Roentgen learned this by accident too. Experimenting to see what effect the rays would have on a photographic film, he wrapped up a film in paper and laid a key on the paper to hold it down. After the film was exposed to the rays and developed he found that he had a perfect outline of the key. The rays had passed through the paper but they would not pass through the dense metal of the key. See Fig. 4.

Two important new fields of research began with Roentgen's accidental discoveries. Doctors used X-ray photographs to study the bones, heart, lungs, stomach and other organs, and physicists began experimenting with the rays to find out what they were. Soon many new discoveries were made about the nature of radiation and its relation to the atom.

Fig. 4. In 1895 Wilhelm Roentgen discovered that he could produce X-rays by passing electricity through a vacuum tube. Parke, Davis and Company.

A few years later Henri Becquerel discovered that the metal, uranium, emitted rays similar to X-rays, and the Curies discovered radium, which was about a million times more active in emitting rays than uranium. These discoveries showed that some sort of reaction was going on in the atoms of certain metals that produced energy and changed the atoms into different metals. Ernest Rutherford was the first to guess that some of the radiation of uranium and radium consisted of particles which have shot out of the nucleus, and that when this happened it changed the nucleus into one of an entirely different element.

Nuclear Reactions

When it was realized that the natural release of particles from certain nuclei could change those nuclei to ones of other types, the next step was to try to change a nucleus from one type to another artificially. One way that this could be done was to bombard a nucleus with high speed nuclear particles that had enough energy to penetrate into the nucleus of the target atom. At that time, shortly after World War I, the only nuclear particles available were those that were emitted from the naturally occurring unstable nuclei, such as radium. These nuclei emitted particles called *alpha particles,* which were actually the nuclei of helium atoms.

In 1919, Rutherford bombarded nitrogen with alpha particles and caused the nitrogen to be transmuted into oxygen. Along with the oxygen a proton was generated, and this proton had more energy than the alpha particle that produced the reaction. This fact showed that some extra energy had been given to the proton, and in fact the extra energy came from the nucleus of nitrogen after the alpha particle had combined with it. Thus, a nuclear reaction with the release of a large amount of energy had been produced artificially for the first time. We see, then, that not only fission and fusion but also certain nuclear reactions can result in the release of energy from the nucleus.

THE FIRST ATOM BOMB

Later, in 1938, two scientists in Nazi Germany, Otto Hahn and Fritz Strassman, succeeded in splitting uranium atoms. Each atom of uranium was split into a barium atom and a krypton atom, and these two atoms together weighed a little less than the uranium atom. The remaining part of the uranium atom was converted into energy according to Einstein's mass-energy formula. The calculations that proved this were made by Lise Meiner, a refugee from Nazi Germany who had fled to Denmark.

The results of this experiment were carried to Einstein who, by this time, had left Germany and gone to Princeton University. A group of American scientists realized it might be possible to build a bomb that would release the tremendous energies available from nuclear fission. They asked Einstein to write to President Roosevelt and explain the terrible consequences that would follow if our enemies were first to build an atomic bomb. Considering the brilliant work that had been done by German scientists in nuclear physics, it

Fig. 5. The atomic pile in which the first controlled release of nuclear energy was achieved on 2 December 1942 at the University of Chicago. Argonne National Laboratory.

seemed likely that Germany would be first to produce such a bomb.

The Manhattan Project

When President Roosevelt read Einstein's letter he decided our scientists should go right to work building an atomic bomb, and started the Manhattan project. This was a gamble since it had not been proved that an atomic bomb was possible. In fact, the Nazi scientists had decided it was not possible and dropped their project. If they hadn't, they might have had the bomb before we did.

The First Chain Reaction

One of the key experiments designed to show whether a bomb could be built was conducted by Enrico Fermi under the stands at Stagg Field, the football field of the University of Chicago. Although it was generally accepted by physicists that splitting the atom would release tremendous energy nobody knew whether such a reaction could be made to continue long enough to be of practical use.

To prove the possibility of controlled nuclear fission, Fermi and the others designed a pile of graphite blocks with uranium rods running through them. According to their calculations, each time a uranium atom in the pile split it would release neutrons that would strike other uranium atoms and cause them to split. Then these atoms would send out more neutrons that would split other atoms and the whole process, which they called a *chain reaction,* would continue as long as the uranium lasted. They believed that if the chain reaction worked and could be controlled then the atom bomb, and also peacetime uses of atomic energy, would be possible. If it didn't work the millions of dollars gambled on the Manhattan project would be wasted.

The physicists designed the pile of graphite blocks according to careful mathematical calculations so that it would support a nuclear fission chain reaction that could be started and stopped at will. Fig. 5 is a drawing of the finished reactor. Since the graphite construction resembled nothing more than a pile of bricks the men working on it started to call it "the pile" and every nuclear reactor since then has been called a pile.

On December 2, 1942 the pile was finished and the test day arrived. Very cautiously, the control rods were pulled out, and the pile became self-sustaining. Further experimenting showed that the reaction could be started and stopped at will. One of the greatest gambles in our nation's history had paid off and the value of the Manhattan project was proved.

QUESTIONS

Check yourself to see whether you remember the information you will need later by completing the following statements. The answers are at the back of the book.

1. The tiny building block of nature that cannot be divided and still keep its basic character is the _____.
2. The theory that all matter is composed of atoms instead of being a continuous substance is called the _____ theory.
3. John Dalton laid the basis for the modern atomic theory by showing how atoms could be _____.
4. A form of matter that contains atoms all of the same type is called an _____.
5. A form of matter composed of two or more different atoms is called a _____.
6. Development of the atomic theory made it possible to discover the principles of _____ _____.
7. Roentgen was the discoverer of _____.
8. Einstein developed the theory that _____ can change to _____.

III. Nature's Engine — The Atom

How can numbers representing very small or large quantities like .0000000134 or 32,000,000,000,000, be written so that they are easy to read? Why does lead stop radiation? Except for the electrons, protons, and neutrons, most of the atom is empty space. How can materials look solid if they are mostly empty space?

To understand radiation and its hazards we must understand the structure of the atom. This knowledge carries the same importance as knowledge of electrical or mechanical principles. An experienced mechanic can handle the ordinary maintenance of machinery; however, if an extraordinary breakdown should occur, the mechanic will most likely have to call in an expert if he lacks the theoretical knowledge. In the same way, knowledge of the atom increases our understanding of radiation, especially its unusual aspects.

Some tradesmen think the structure of the atom is too complicated for them to understand. This is nonsense! Anyone intelligent enough to learn a trade has more than enough mental ability to understand the atom. Most skilled workmen could explain what the pistons, crankshaft, valve lifter and cylinders of an automobile engine are and how they operate together. As Fig. 1 shows, an automobile engine has many kinds of working parts and its operation is quite complicated. Nature's engine, the atom, has only three main kinds of working parts: electrons, protons, and neutrons.

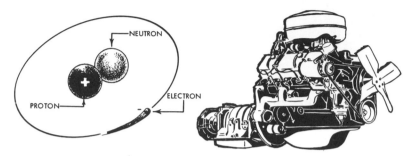

Fig. 1. An automobile engine has many kinds of working parts. An atom has only three main kinds: electrons, protons and neutrons.

The study of the structure of the atom is not completed and probably never will be. Knowledge will continue to increase and change. Still, anyone who understands the material in this chapter will have enough fundamental knowledge to enable him to understand the new developments as they come along.

POWERS OF TEN

Before we go into the structure of the atom, let's take time to learn something that will make it much easier to talk about the atom and also to work everyday math problems. Measurements of the parts of the atom are much like measurements made in astronomy in that many of the figures used are so large or so small that they are hard to write and to handle. To simplify figures like .0000000134 and 32,000,000,000,000 we use a shorthand system that eliminates writing down all the zeros. For example, instead of writing 100 we write $1 \times 10 \times 10 = 1 \times 10^2$. This is read *one times ten to the second power*.

In this notation, 400,000,000,000 would be written 4×10^{11} and read *four times ten to the eleventh power*. The number above the ten gives the number of places you have to move the decimal point in 400,000,000,000 to get 4. If you did not know about the powers of ten you might read the number 400,000,000,000 as *four with eleven zeros*. Writing it as 4×10^{11} is just a scientific way of saying the same thing.

TABLE I. POWERS OF TEN		
1×10^4 =	10,000	= 10 x 10 x 10 x 10. (Read ten to the fourth power)
1×10^3 =	1,000	= 10 x 10 x 10. (Read ten to the third power or ten cubed)
1×10^2 =	100	= 10 x 10. (Read ten to the second power or ten squared)
1×10^1 =	10	= 10. (Read ten to the first power)
1×10^0 =	1	= 1. (Read ten to the zero power)
1×10^{-1} =	.1	= 1/10. (Read ten to the minus first power)
1×10^{-2} =	.01	= 1/(10 x 10) or 1/100. (Read ten to the minus second power)
1×10^{-3} =	.001	= 1/(10 x 10 x 10) or 1/1000. (Read ten to the minus third power)
1×10^{-4} =	.0001	= 1/(10 x 10 x 10 x 10) or 1/10,000. (Read ten to the minus fourth power)

Moving the Decimal Point

Now let's go a step further. Since 10^2 means *move the decimal point two places to the right,* it is logical that 10^{-2} would mean the opposite, *move the decimal two places to the left.* When a number is written with no decimal point it is understood that the decimal point comes at the end. Then 1×10^2 means 100 and 1×10^{-2} means .01. If you remember that the minus sign always means a smaller number, you should be able to remember which way to move the decimal. Table I is a list of the powers of ten. These numbers . . . 10^{-4}, 10^{-3}, 10^{-2}, 10^{-1}, 10^0, 10^1, 10^2, 10^3, 10^4 . . . can get large enough or small enough for any purpose and each one is exactly ten times as large as the one before. The numbers above and to the right of the 10's are called *exponents.*

The method is just the same for more complicated numbers, such as 3820000000 or .01304. The *standard form* of any number is a power of ten multiplied by a number between 1 and 10, including 1. The best way to find the standard form of a number is to move the decimal point until you have a number between 1 and 10. For an example, let's take the number 3820000000. Each time the decimal point is moved one place to the left this has the same effect as dividing out one 10. It works like this: $3820000000 = 382000000.0 \times 10 = 38200000.00 \times 10^2 = 3820000.000 \times 10^3 = \ldots = 3.820000000 \times 10^9$. To get 3.82, the number between 1 and 10, out of the original number, you have to move the decimal to the left nine places, so *3.82 $\times$ 10⁹* is the final form.

To change .01304 to a number between 1 and 10 you would have to move the decimal two places to the right, so the standard form of this number is 1.304×10^{-2}. The reason we use numbers from between 1 and 10, including 1, is that multiplication and division are simpler if all the numbers involved are written in the same form.

There is nothing complicated about using powers of ten. It is simply a more convenient way to write very large or very small numbers. All you need to remember is that the exponent tells you which way and how far to move the decimal. An exponent without a minus sign means move the decimal to the right. An exponent with a minus sign means move the decimal to the left.

To find out whether you understand, try working the following problems and then check your answers with those at the end of the book. If you are having trouble, read the chapter over or try to get someone to explain it to you. Many times talking something over with another student clears up the problem.

PROBLEMS

1. Write these numbers in the usual way, using zeros:
    (a) 2×10^5 (b) 5×10^{-4} (c) 6.7×10^{-6}
    (d) 8.54×10^6 (e) 34×10^7 (f) 142×10^4
2. Write these numbers as powers of ten multiplied by numbers between 1 and 10, using 1 if necessary but not 10:
    (a) .0034   (b) 1003   (c) 4520000   (d) .000103
3. The diameter of one hydrogen atom is 1.35×10^{-8} centimeters. Write this number without using the power of ten.
4. If one pound of matter were completely converted into energy it would produce 10,000,000,000,000,000 calories. Write this figure as one times a power of ten.
5. Write the correct exponent in place of the question mark:
$$25,470,000,000,000,000,000 = 2547 \times 10? = 254.7 \times 10?$$
$$= 25.47 \times 10? = 2.547 \times 10?$$

Multiplying and Dividing

Now let's go one step further and learn to multiply and divide numbers written this way. Powers of ten are very useful in all our work because they make multiplication and

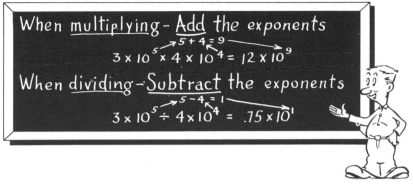

Fig. 2. Multiplying and dividing with powers of ten.

division much easier. Also, if you use a slide rule, they show exactly where the decimal point goes in your answer.

To multiply, simply multiply the simple numbers and *add* the exponents. For instance, $100 \times 100 = 10,000$ when it is written in the usual way. In scientific notation this problem is written $1 \times 10^2 \times 1 \times 10^2 = 1 \times 10^4$. To multiply 28.3 by 30,000,000,000 write the numbers as 2.83×10^1 and 3×10^{10}. Then $2.83 \times 10^1 \times 3 \times 10^{10} = 8.49 \times 10^{11}$. We got this answer by multiplying 2.83 by 3 and adding the exponents 1 and 10.

You can multiply three or more numbers the same way; by multiplying all the small numbers together and adding all the exponents. For example:

$$28.3 \times 30,000,000,000 \times 30,000,000$$
$$= 2.83 \times 10^1 \times 3 \times 10^{10} \times 3 \times 10^7$$
$$= (2.83 \times 3 \times 3) \times (10^1 \times 10^{10} \times 10^7)$$
$$= 25.47 \times 10^{18}$$

It would be practically impossible to multiply these numbers in the usual way without getting confused about how many zeros there should be.

This works just as well for very small numbers. Multiplying .00013 by .00432 we have $1.3 \times 10^{-4} \times 4.32 \times 10^{-3} = 5.616 \times 10^{-7}$. Multiplying .00013 by 4320 we have $1.3 \times 10^{-4} \times 4.32 \times 10^3 = 5.616 \times 10^{-1}$ since, when -4 and 3 are added, the answer is -1.

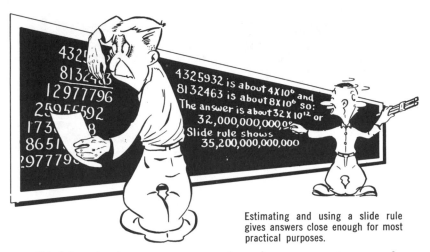

Estimating and using a slide rule gives answers close enough for most practical purposes.

Division is the same except that the exponents are subtracted instead of being added and the simple numbers are divided. For example, 45,000,000 divided by 15,000 is written as $(4.5 \times 10^7) \div (1.5 \times 10^4) = 4.5/1.5 \times 10^3$ or 3×10^3. We can divide 45,000,000 by .0015 by writing $4.5 \times 10^7 \div 1.5 \times 10^{-3} = \dfrac{4.5}{1.5} \times 10^{10} = 3 \times 10^{10}$. In this case we need to remember that subtracting -3 from $+7$ gives the answer $+10$, since $+7 - (-3) = +7 + 3 = +10$. Fig. 2 shows how this way of multiplying and dividing works.

Estimating Answers

Understanding how to multiply and divide with powers of ten is especially valuable to any tradesman or technician for estimating approximate answers or for figuring where the decimal place goes in a slide rule problem. For example, look at this problem: $28,793 \times 403,107$. This is a complicated multiplication problem and most people would not be sure they were right after they got through with it. However, by writing down approximate numbers with powers of ten, we can find the approximate answer and the decimal point. The number, 28,793 is changed to 3×10^4 because it is closer to 30,000 than to any other round number and 403,107 is changed to 4×10^5 for the same reason. Then

$3 \times 10^4 \times 4 \times 10^5 = 12 \times 10^9$ or 1.2×10^{10}. This means the answer should be close to 1.2×10^{10}, or 12,000,000,000.

The actual answer 11,606,659,851 which shows that our approximation was correct. Since we made the approximation before and know that the answer should be around 12 billion we can have confidence that our more accurate calculation is correct and that the decimal point is in the right place.

This is a good example of how science and technology are dependent on each other for help. The craftsman who can borrow scientific methods like this and adapt them to his work will be head and shoulders above the average.

PROBLEMS

To see whether you understand this method of multiplying and dividing, work the following problems and check your answers with the ones given at the end of the book.

1. Multiply:
   (a) 1.8×10^{21} by 1.6×10^4 (b) 1.8×10^{21} by 1.6×10^{-4}
   (c) 1.8×10^{-21} by 1.6×10^{-4}

2. Divide:
   (a) 4.8×10^{21} by 1.6×10^4 (b) 4.8×10^{21} by 1.6×10^{-4}
   (c) 4.8×10^{-21} by 1.6×10^{-4}

3. Find the approximate answers by estimating:
   (a) $39,642 \times 120,899$ (b) $39,642 \times .00120899$
   (c) $.39642 \times .00120899$

4. Find the approximate answers by estimating:
   (a) $39,642 \div 120,899$ (b) $39,642 \div .00120899$
   (c) $.39642 \div .00120899$

5. Find the exact answers:
   (a) 23000×3010 (b) $23000 \times .00301$ (c) $.023 \times .00301$

6. Find the exact answers:
   (a) $739500 \div 1020$ (b) $739500 \div .00102$ (c) $.07395 \div .00102$

THE WORKING PARTS OF THE ATOM

The parts of the atom, as we said before, are no more complex than the parts of the auto engine. In this section we will consider the three main parts of the atom—electrons, neutrons, and protons. Scientists have discovered other very small particles of matter, but most of these exist for only extremely short times once they are removed from the atom.

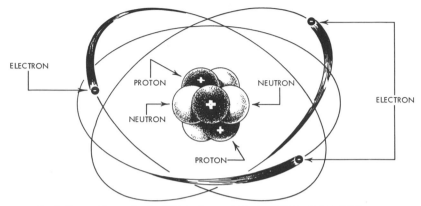

Fig. 3. The working parts of a lithium atom—electrons, protons, and neutrons.

The Atom

An atom consists of a heavy, dense center, called the nucleus, surrounded by electrons which spin around it at tremendous speeds, as shown in Fig. 3. The nucleus is a cluster of two kinds of particles called *protons* and *neutrons,* which are practically the same size and weight. The difference between them is that a proton has a positive charge of electricity and a neutron has no electric charge. This should be easy to remember since the beginning of the word neutron gives a clue to its meaning. The *neutron* is electrically *neu*tral. The mass of a neutron or proton is about 1.7×10^{-24} grams. If you remember our discussion of powers of ten you can see that this is a very small number.

Electrons. Electrons are negatively charged particles of matter that orbit around the nucleus much the way a satellite orbits around the earth. Electrons are much lighter than neutrons or protons—the mass of an electron is 9.1×10^{-28} grams, only about 1/1840 the mass of a proton or neutron. The atom itself, which is 10,000 times larger than the nucleus, is still so small that there are six sextillion (6×10^{21}) atoms in a drop of water.

The negative electric charge of an electron is exactly as strong as the positive charge of a proton. When the atom is in its normal state, the number of negatively charged electrons orbiting around the nucleus is the same as the number of protons in the nucleus. Since each electron neutralizes the charge of one proton, the atom as a whole has no electric charge.

Matter and Space

It is hard to imagine how small the parts of the atom are. Atoms are made up mostly of space and everything in the world, including ourselves, contains much more space than matter. When you say that someone is full of hot air, you are pretty close to the truth!

To get an idea of how much space there is in an atom, imagine that the nucleus of a hydrogen atom could be enlarged to the size of a baseball, as shown in Fig. 4. Then the closest electron to the nucleus would be eight blocks away. Scientists have figured that if you could take all the space out of a 200 pound man and pack the neutrons, protons, and electrons together, the amount of condensed material remaining would be no larger than a speck of dust. This description of matter as being mostly empty space is hard to grasp, but the fact is, pure, condensed matter is unbelievably dense and heavy. A ball of pure, condensed matter the size of a drop of water would weigh two million tons!

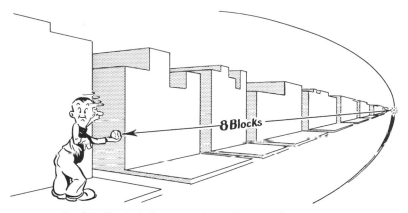

Fig. 4. Matter, including atoms, is mostly space. If the nucleus of an atom could be enlarged to the size of a baseball, the closest electron would be eight city blocks away.

It is easier to visualize this if you remember that all these distances are extremely small. We are in the same position as if we were looking at something from a great distance. For instance, if we were looking at a busy highway at night from a long way off, we would see the car lights, but they would look like one continuous strip of light. As we got closer to the lights we would see that they were individual moving lights, but there would seem to be only one light for each car. Coming closer still we would see that each car is a source of at least two lights and that there is a considerable amount of space between the lights even though at a distance they looked like one continuous strip of light.

Shielding

It is important to know about atomic structure because a number of important properties of radiation depend on the interaction of radiation with the atom. For example, lead is a good shielding material for most radiations because there are more electrons in an atom of lead than in most light atoms. This means, rays trying to pass through lead are likely to hit particles of matter and be stopped instead of passing through the space between the particles. A light material such as aluminum has fewer electrons, so rays have a better chance of passing through without being stopped.

ATOMIC NUMBER AND MASS NUMBER

As you learn more about the atom, you will see that nuclear energy—as the name implies—is concerned primarily with changes in the nucleus of the atom. The *atomic number* and the *mass number* of an atom are two numbers that tell the contents of the nucleus.

The atomic number of an atom is the number of protons in the nucleus of the atom. The atoms found in nature have from one to 92 protons. Atoms with from 93 to 105 protons, which don't exist in nature, have been created by scientists in particle accelerators.

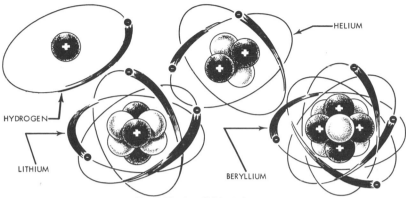

Fig. 5. The four lightest elements.

The number of protons indirectly determines the chemical nature of the atom, which means its ability or inability to form compounds with other atoms; an atom with one proton is hydrogen, one with two protons is helium, one with three protons is lithium, and so on. See Fig. 5. There is one electron for each proton and the number and arrangement of the electrons governs the chemical properties of an atom. If the number of protons in an atom changes it turns into an atom of an entirely different element. For example, when an atom of the metal, radium, loses two protons it becomes radon, a gas.

The mass number of an atom is the total number of protons plus neutrons in the nucleus. The actual weight of an atom depends both on the mass number and on how the protons and neutrons are held together in the nucleus.

Isotopes

Most elements have several different kinds of atoms. If the number of protons in the nucleus of an atom changes, the atom turns into a different element, but if a neutron is added or taken away nothing happens except that the mass number is increased or decreased by one. These different forms of atoms of the same element are called *isotopes* of that element. Two isotopes of the same element have the same atomic

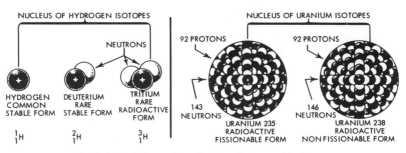

Fig. 6. At the left are shown the three isotopes of hydrogen; at right, two isotopes of uranium.

number but different mass numbers and the difference is caused by variation in the number of neutrons.

The simplest example of this is hydrogen. Fig. 6 shows the three isotopes of hydrogen, ^1_1H, ^2_1H, and ^3_1H, and two isotopes of uranium. Most hydrogen atoms have no neutrons. They consist of one proton and one electron and their mass number is 1. A few have one neutron and a very small number have 2. Those with a nucleus of one proton and one neutron in the nucleus are about twice as heavy as the first kind. Their mass number is 2 but their chemical properties are just the same as those of ordinary hydrogen. When these heavy atoms combine with oxygen, they form water which is much heavier than ordinary water. The third hydrogen isotope has an atomic weight of 3 and is radioactive.

Atomic Weight

As mentioned previously, most elements have several isotopic forms, and among these isotopes a few usually are stable; that is, they are not radioactive. The three isotopes of hydrogen were listed previously; of these, ^1_1H and ^2_1H are stable and they are found in nature. But these two are not found in nature in equal amounts. Natural hydrogen is always composed of 99.985 percent ^1_1H and 0.015 percent ^2_1H. The same holds for the element carbon *(C)*, which is found in nature as 98.89 percent of the isotope $^{12}_6C$ and 1.11

percent of the isotope $^{13}_{6}C$. Another example is iron (*Fe*). which has four stable isotopes in the proportions 91.6 percent of $^{56}_{26}Fe$, 5.9 percent $^{54}_{26}Fe$, 2.2 percent of $^{57}_{26}Fe$ and 0.33 percent of $^{58}_{26}Fe$. Since each stable isotope of a given element has a definite number of protons, neutrons, and electrons, each isotope will have a certain weight, and each weight will be different. Since all isotopes of a given element have the same chemical properties, however, it is convenient to deal with an average weight for all of them. The average is taken in such a way that the most common isotope (the one present in highest percentage) makes the biggest contribution to the average.

Another convenience is to consider the weight not of one average atom of an element but of a standard number of atoms of that element (containing the natural mixture of stable isotopes). This standard number is called *Avogadro's number* and is equal to 6×10^{23}. Then the *atomic weight* of any element is defined as the weight, in grams, of 6×10^{23} atoms of that element. The number 6×10^{23} is defined as the number of atoms in exactly 12 grams of the most common isotope of carbon, $^{12}_{6}C$. The reference for atomic weights used to be 16 grams of oxygen, but the new standard is carbon. Table II on the following page lists the atomic weights of the elements. A complete listing of the elements shown in related groups is called a *periodic table*.

Finding the Number of Neutrons

The number of isotopes of different elements varies considerably—tin has twenty-five—and most of them are radioactive. Both of the two isotopes of uranium shown in Fig. 6 have 92 protons and 92 electrons, because the atomic number of uranium is 92. But $^{235}_{92}U$ has 143 neutrons and $^{238}_{92}U$, the heavier isotope, has 146. Both of these are radioactive but only $^{235}_{92}U$ atoms can fission on capturing a neutron.

TABLE II. THE ELEMENTS

Atomic Number	Name of Element	Symbol of Element	Atomic Weight	Atomic Number	Name of Element	Symbol of Element	Atomic Weight
1	Hydrogen	H	1	53	Iodine	I	127
2	Helium	He	4	54	Xenon	Xe	131
3	Lithium	Li	7	55	Cesium	Cs	133
4	Beryllium	Be	9	56	Barium	Ba	137
5	Boron	B	11	57	Lanthanum	La	139
6	Carbon	C	12	58	Cerium	Ce	140
7	Nitrogen	N	14	59	Praseodymium	Pr	141
8	Oxygen	O	16	60	Neodymium	Nd	144
9	Fluorine	F	19	61	Promethium	Pm	147
10	Neon	Ne	20	62	Samarium	Sm	150
11	Sodium	Na	22	63	Europium	Eu	152
12	Magnesium	Mg	24	64	Gadolinium	Gd	157
13	Aluminum	Al	27	65	Terbium	Tb	159
14	Silicon	Si	28	66	Dysprosium	Dy	162
15	Phosphorus	P	31	67	Hilmium	Ho	165
16	Sulfur	S	32	68	Erbium	Er	167
17	Chlorine	Cl	35	69	Thulium	Tm	169
18	Argon	A	39	70	Ytterbium	Yb	173
19	Potassium	K	39	71	Lutecium	Lu	175
20	Calcium	Ca	40	72	Hafnium	Hf	179
21	Scandium	Sc	45	73	Tantalum	Ta	181
22	Titanium	Ti	48	74	Tungsten	W	184
23	Vanadium	V	51	75	Rhenium	Re	186
24	Chromium	Cr	52	76	Osmium	Os	190
25	Manganese	Mn	55	77	Iridium	Ir	193
26	Iron	Fe	56	78	Platinum	Pt	195
27	Cobalt	Co	59	79	Gold	Au	197
28	Nickel	Ni	59	80	Mercury	Hg	201
29	Copper	Cu	64	81	Thallium	Tl	204
30	Zinc	Zn	65	82	Lead	Pb	207
31	Gallium	Ga	70	83	Bismuth	Bi	209
32	Germanium	Ge	73	84	Polonium	Po	210
33	Arsenic	As	75	85	Astatine	At	211
34	Selenium	Se	79	86	Radon	Rn	222
35	Bromine	Br	80	87	Francium	Fr	223
36	Krypton	Kr	84	88	Radium	Ra	226
37	Rubidium	Rb	85	89	Actinium	Ac	227
38	Strontium	Sr	88	90	Thorium	Th	232
39	Yttrium	Y	89	91	Protactinium	Pa	231
40	Zirconium	Zr	91	92	Uranium	U	238
41	Columbium	Cb	93	93	Neptunium	Np	239
42	Molybdenum	Mo	96	94	Plutonium	Pu	239
43	Technetium	Tc	99	95	Americium	Am	241
44	Ruthenium	Ru	102	96	Curium	Cm	242
45	Rhodium	Rh	103	97	Berkelium	Bk	245
46	Palladium	Pd	107	98	Californium	Cf	246
47	Silver	Ag	108	99	Einsteinium	E	253
48	Cadmium	Cd	112	100	Fermium	Fm	256
49	Indium	In	115	101	Mendelevium	Mv	256
50	Tin	Sn	119	102	Nobelium	No	254
51	Antimony	Sb	122	103	Lawrencium	Lw	257
52	Tellurium	Te	128	104 & 105	Being Studied		

NOTE: Elements 1 through 92 occur normally in nature. Elements 93 and above are those discovered by man as a result of transmutation.

The number of neutrons in an atom of any isotope can be found by subtracting the atomic number from the mass number, since the atomic number is the number of protons and the mass number is the total number of neutrons and protons in the nucleus. To see how the symbols for isotopes are interpreted, let's take the case of potassium. The letter K stands for potassium. In the isotope of potassium, $_{19}^{39}$K, the mass number is 39 and the atomic number is 19. This means there are 19 protons and 19 electrons, and the number of neutrons is 20 since $39 - 19 = 20$.

SUMMARY

When numbers like 234,000,000 and .000000234 are written as powers of ten, the number above the 10 shows how many places to move the decimal point to get the number in its usual form. If the number above the 1.0 has a minus sign, the decimal point is moved to the left. If it doesn't have a minus sign, the decimal is moved to the right. To multiply numbers written with powers of ten, we add the numbers appearing above the 10's. To divide, we subtract one from the other.

The atom consists of a *nucleus* with negatively charged *electrons* orbiting around it. The nucleus is a cluster of positively charged *protons* and electrically neutral *neutrons*.

The *atomic number* of an element is the number of protons in the nucleus of each atom. Each element has a different atomic number and all isotopes of the same element have the same atomic number.

The *mass number* of an isotope is the total number of neutrons and protons in the nucleus. The number of neutrons in the nucleus of an isotope of any element is the mass number minus the atomic number.

Atoms having the same atomic number but different mass numbers are called *isotopes*. The only difference between isotopes of the same element is the number of neutrons in the nucleus.

QUESTIONS

To test your understanding of the structure of the atom, try to complete these statements. Then check your answers with those given at the end of the book.

1. The parts of the atom that orbit around the nucleus are the _____.

2. The part of the atom with a negative electrical charge is the _____.

3. The center of the atom is the _____.

4. The part of the atom that has a positive charge is the _____.

5. The nucleus contains _____ and _____.

6. The atomic number tells the number of _____ in the nucleus.

7. To find the number of (a)_____ in the nucleus of an isotope, subtract the atomic (b)_____ from the (c)_____ number.

8. The symbol for zinc is Zn. Fill in the following facts about the zinc isotope $^{65}_{30}$Zn.
   mass number (a)_____; atomic number (b)_____; number of protons (c)_____; number of neutrons (d)_____; number of electrons (e)_____.

IV. Harnessing Atomic Power

Where does the energy of the sun and stars come from? When a log splits no energy is created. Why should the splitting of an atom's nucleus create energy? Why will a fission chain reaction start in a cube of uranium and not in a flat plate of exactly the same weight? How is atomic power converted to electricity?

Now that you have some understanding of atomic structure we can begin to talk about nuclear power in more detail. As you read in Chapter 1, when a heavy atom splits into two light atoms this process is called nuclear fission. Fusion is the opposite process—the nuclei of several light atoms are forced together to make one heavier atom.

Fission and fusion are always going on in nature. Fusion is the source of the energy of the sun and stars. Fig. 1 is a photograph of fusion activity on the sun. Light, heat, and

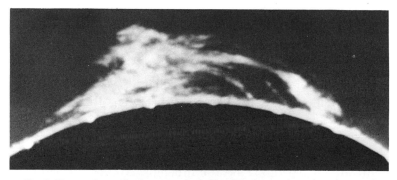

Fig. 1. The sun is an atomic power plant. The disturbance shown here is caused by violent fusion activity. Mount Wilson and Palomar Observatories.

other forms of energy are released when hydrogen atoms in the sun combine and become helium atoms. In some stars much hotter than the sun, helium atoms constantly combine to become carbon. In stars still hotter than this, carbon atoms combine to form still heavier atoms and from then on, a combination of fusion, fission, and radioactivity produces all the elements that exist. These different kinds of reactions are occurring in stars all over the universe.

In any sample of uranium, certain atoms are spontaneously fissioning all the time. This is because the atoms of the rare isotope of uranium, ^{235}U, split easily, and some of these are always mixed in with more stable atoms of the common isotope, ^{238}U.

When fission goes on in a nuclear reactor, large amounts of heat are produced at a constant rate, and this heat can be converted to electricity by mechanical means. The fission process also changes some materials in the reactor to radioisotopes. Since 1942, when the first pile was built, nuclear reactors have been constructed all over the country. They supply electric power and radioisotopes for use in medicine, agriculture, and industry. No fusion reactors have been built yet since scientists are still working on this problem.

ENERGY IN THE ATOM

Binding Energy

As you learned before, the nucleus of an atom is made up of neutrons and protons. Each proton has a positive electric charge. Neutrons have no electric charge. You are probably wondering now how the protons can stay together since positively charged objects repel each other.

To understand this we have to remember that mass can change to energy and back again. Part of the mass of the neutrons and protons of the nucleus is transformed into energy, called *binding energy,* which keeps them from flying apart. For example, a helium nucleus, consisting of two protons and two neutrons, weighs less than the protons and neu-

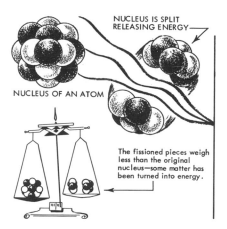

NUCLEUS IS SPLIT
RELEASING ENERGY

NUCLEUS OF AN ATOM

The fissioned pieces weigh
less than the original
nucleus—some matter has
been turned into energy.

Fig. 2. Part of the mass of the nucleus is transformed into energy.

trons weigh separately outside the nucleus. Some of the mass of the particles has changed to binding energy which holds the helium nucleus together.

Why Fission Releases Energy

In the 1930's, scientists were bombarding atoms in machines called *particle accelerators,* or *"atom smashers."* An accelerator can shoot a stream of high speed particles at a target of any matcrial, such as uranium. When a neutron moving at the right speed strikes the nucleus of a ^{235}U atom, it penetrates into the nucleus and makes it very unstable. This unstable nucleus splits into two nuclei of lighter atoms and a number of free neutrons. Since the binding energy needed to keep these lighter nuclei together is less than the binding energy of the original ^{235}U atom, there is extra energy present. This energy does not turn back into mass again. It remains energy and is released in the form of heat and radiation. See Fig. 2.

THE CHAIN REACTION

During World War II, while we worked on splitting the uranium atom so as to release huge amounts of energy, the Germans were working on the same project. Our first goal was

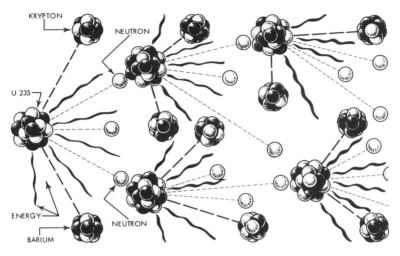

Fig. 3. A fission chain reaction.

to develop the atom bomb before they did. But before this could be done it was necessary to prove that the process of nuclear fission could continue automatically after it was set off. Before 1942 scientists had been able to split the atom only under laboratory conditions. It took more energy to make the atoms split than the fission released and the process stopped immediately unless more energy was fed in.

The scientists thought that if conditions were right some of the neutrons that escaped from the nucleus when a uranium atom was split would hit other ^{235}U atoms and cause them to fission too, as shown in Fig. 3. If an average of one neutron from every split atom struck and fissioned another atom, the reaction could continue indefinitely and stay under control. But if more than one neutron per fission hit and split another atom the reaction would go out of control, releasing so much energy all at once that it would be an atomic explosion. The process of fission continuing automatically was called a *chain reaction* because each step would lead to the next step and carry the reaction along.

The Multiplication Factor

You are probably aware of how fast something will grow if

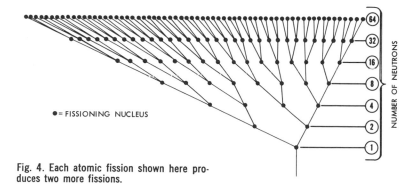

NUMBER OF NEUTRONS

● = FISSIONING NUCLEUS

Fig. 4. Each atomic fission shown here produces two more fissions.

it is doubled at regular intervals. You may have heard the story of the man who went to work for a penny a day and agreed to have his wages doubled every day for a month. On the second day he got two cents; on the third day he got four cents; and on the fourth day he got eight cents, which seems like a bad bargain. But if you figure that the sum is doubled every day for thirty days you see that on the last day alone he received $5,368,709.12.

The chain reaction in the atom bomb works the same way, as Fig. 4 shows. If two neutrons from a split atom hit two more atoms these atoms split and release a total of four neutrons and so on. You can see from the drawing how fast this would multiply, and in the bomb it happens almost instantaneously.

Mousetraps and Ping Pong Balls

One of the best examples of the chain reaction has been shown on television and in classrooms everywhere. In this demonstration a large number of mousetraps are set in a small area close together. A ping pong ball is balanced on the arm of each trap so that the ball will fly when the trap is sprung. The traps represent atoms and the ping pong balls neutrons.

When one ping pong ball is dropped over the traps the result is instant pandemonium with balls flying in all directions. Each ball from a sprung trap flies off and bounces around, setting off one or more additional traps. The whole thing is over in a matter of seconds with hundreds of balls up in the air at once.

This is an excellent example because it demonstrates the predictable nature of the chain reaction as a whole, in spite of the random action and unpredictability of individual atoms and neutrons. When the first ball is dropped no one can tell whether it will set off one trap or three, but it is almost certain to set off at least one. As each trap is sprung, there is no way of knowing whether that particular ball will fly out of the area of traps and not set off any, or whether it will bounce around and set off three or four traps. All we can predict is that after the first ball is dropped there will be enough traps sprung to start and continue the action until most or all of the traps are sprung.

This is just what happens in a fission chain reaction. The splitting of one atom releases several neutrons but there is no way of knowing what these neutrons will do. Some of them will pass through the surface of the metal and escape. Others will bounce around without ever hitting a nucleus. Still, by setting up the right conditions, scientists can predict about how many neutrons will be flying around at any moment, and how many atoms will be split.

Criticality, Subcriticality, Supercriticality

As long as, on the average, each atom that splits causes one other atom to split, the reaction will continue but it will not grow. When this happens the reaction is called *critical*. But if the average number is more than one, the reaction will go out of control and this situation is called *supercritical*. If the average number of fissions caused by one atom splitting is less than one the reaction will die out. This is called the *subcritical* condition.

In other words when the average number of fission-producing neutrons is less than one per fission, the reaction is *subcritical;* when the average is one, the reaction is *critical;* and when the average is more than one, the reaction is *supercritical.*

CRITICAL CONDITIONS

The size and shape of the fuel and moderator elements of a reactor are an important part of its design. This is because their dimensions and positions determine the directions of the moving neutrons and the average number of neutrons at any point. Many of them will go through the surface of the reactor to the outside.

Critical Size

One way to make sure enough neutrons will stay inside the reactor to sustain the chain reaction is to make the reactor large enough. If the fuel elements are smaller than a certain size, called the *critical mass,* so many neutrons will escape that the reaction can't continue. Increasing the size of the fuel element to the critical mass makes a continuing reaction possible.

Reflectors. A fuel element smaller than the critical size can sustain a chain reaction if it is surrounded by a reflector so that the neutrons escaping from the surface of the fuel can hit atoms of the reflector and bounce back in. Many reactors are made with reflectors which conserve neutrons that would otherwise be lost. Ordinary water and light elements such as beryllium make good reflectors.

Size and Shape

Changing the shape of a critical mass can make it *subcritical.* As Fig. 5 shows, a flat plate of uranium has more surface area than a cube of the same weight. This means more neu-

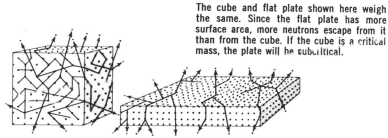

The cube and flat plate shown here weigh the same. Since the flat plate has more surface area, more neutrons escape from it than from the cube. If the cube is a critical mass, the plate will be subcritical.

Fig. 5. The shape of a critical mass affects the chain reaction.

trons will escape from the surface. If the cube is just large enough to sustain a chain reaction, the flat plate of the same weight will not sustain one because there will not be enough neutrons splitting atoms inside the plate.

Furnaces or Bombs

The difference between an atomic bomb and an atomic reactor is that in a bomb the fission chain reaction goes supercritical all at once and maintains supercriticality for as long as possible. It is the same as the difference between lighting a firecracker at the fuse so that the powder goes off all at once, and opening up the firecracker and lighting the loose powder so that it burns slowly. A reactor produces the same heat and other energy as a bomb, but it does this much more slowly and stays under control. For this reason reactors are sometimes called atomic furnaces.

It is practically impossible for a reactor to become a bomb by accident. An atomic bomb is specially designed so that pieces of fuel suddenly brought together achieve supercriticality at the same instant, before the bomb can melt down or blow apart without much fission occurring. Reactors are constructed differently. In a reactor any excessive heat would slow down the reaction, and supercriticality would produce so much heat that the fuel would melt in an instant. This would change the shape of the fuel element so much that the reaction would stop altogether. Besides, all reactors are equipped with automatic safety devices which begin to function as soon as emergency conditions are detected.

NUCLEAR REACTORS

In a nuclear reactor, sometimes called an atomic pile or atomic furnace, a fission chain reaction goes on at a constant rate. This means an average of one neutron per split atom hits another atom and makes it split. You don't have to do anything to a nuclear reactor to make it operate. The spontaneous fission of ^{235}U that goes on all the time will start the chain reaction when all the parts are in place.

Fuel

Natural uranium is made up mostly of atoms of the isotope ^{238}U. These ^{238}U atoms will not split unless they are hit by fast-moving neutrons and even then no chain reaction occurs. Just the opposite holds for ^{235}U, that is, slow neutrons are absorbed and result in fission; also a fission chain reaction can be maintained. One of the fuels used in reactors is enriched uranium, which is uranium that has been processed so that it will have many more atoms of the rare isotope ^{235}U than natural uranium. The other fuel is plutonium, an element produced in reactors and particle accelerators.

Moderators

As mentioned, the easiest way to split a ^{235}U atom is to hit it with a neutron moving slowly enough so that it can be absorbed into the nucleus. Most of the neutrons released when this atom fissions are going too fast to cause fission in another ^{235}U atom. But if these neutrons pass through a material such as water, heavy water, beryllium, or graphite, they bounce around and are slowed down enough so that they have a better chance of causing fission when they hit the atom. The material used to slow the neutrons is called the *moderator*.

As you read in an earlier chapter, the first atomic pile, built in 1942, was constructed of graphite blocks with uranium rods running through them. The blocks were arranged so that the

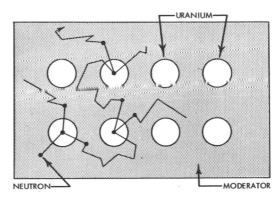

Fig. 6. The moderator shown here reduces the speed of the neutrons.

uranium rods formed a lattice completely enclosed by graphite. Fig. 6 shows what happens in a pile of this type. Fast moving neutrons released by fission of uranium atoms hit atoms in the graphite moderator. By the time one of these neutrons enters another uranium rod, these collisions have reduced its speed enough so that it has a good chance of being absorbed by a ^{235}U atom in the fuel rod and causing fission.

Control Rods

Atoms of certain materials, such as cadmium and boron, will absorb neutrons without splitting. Rods made of these materials are used in reactors to control the chain reaction and keep it from becoming supercritical. If the average number of neutrons splitting ^{235}U atoms becomes higher than one per fission, the control rods are pushed further into the moderator where they can slow down the reaction by absorbing some of the neutrons. Control rods are pulled out to start a reactor and dropped in to shut it down. Fig. 7 shows how the fuel elements, moderator, and control rods function together in a reactor.

Shielding

Fig. 8 is a photograph of a reactor and Fig. 9 is a drawing

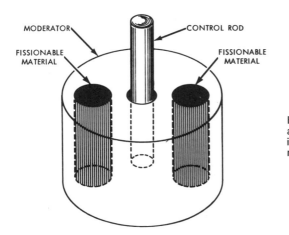

MODERATOR

CONTROL ROD

FISSIONABLE
MATERIAL

FISSIONABLE
MATERIAL

Fig. 7. Control rod in a reactor for starting, regulating, and stopping chain reaction.

Fig. 8. A reactor at Oak Ridge Research Laboratories.

showing its essential parts. Much of the bulk of this reactor is
in its walls since the shielding must be at least eight feet thick
in order to stop the radiation.

As you have seen, a nuclear reactor must have four parts to
operate: 1) a moderator to slow the neutrons down, 2) a fuel
element large enough so that there will be at least one neutron per fission splitting a new atom or smaller fuel element

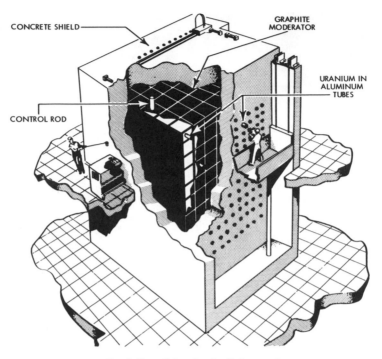

Fig. 9. Essential parts of a fission reactor.

and a reflector to bounce some of the escaping neutrons back in, 3) control rods to keep the reaction from going super-critical, and 4) shielding to stop the neutrons and the other radiation resulting from fission. In addition, the fuel elements must be the right shape—in other words, there must be a large enough volume compared to the surface area so that there will not be too many neutrons lost.

USES OF ATOMIC REACTORS

Producing Electric Power

Fig. 10 shows a cutaway diagram of the Dresden nuclear power station near Chicago. Many such reactors are in use throughout the world and more will be built as research makes them cheaper to build and operate. There are several

Fig. 10. Dresden Nuclear Power Plant, near Chicago. Commonwealth Edison Co.

different types of nuclear power plants but the basic principles are the same: they all use the heat produced by fission in the reactor to change water into steam. Then the steam operates the turbine-generator that produces electricity.

Radioisotope Production

Many elements become radioactive if they are bombarded with neutrons. These materials can be made into radioisotopes if they are left in a reactor for some definite period of time, since in a reactor free neutrons are moving around in all directions. This is the most important radioisotope production method in use now.

Breeder Reactors

The common isotope of uranium, ^{238}U, can be changed in nuclear reactors to another element, plutonium. Plutonium can sustain a fission chain reaction and makes an excellent fuel for reactors. Reactors in which this is done are called *breeder* reactors. They manufacture more fuel than they consume and they can produce electric power and radioisotopes at the same time.

FUSION

If fusion reactors, Fig. 11, could be built they would have several advantages over fission reactors. Fusion releases much more heat than fission and it is easier to control. The hydrogen, water, or other hydrogen compounds they would use for fuel are much more plentiful than the uranium materials used in fission reactors.

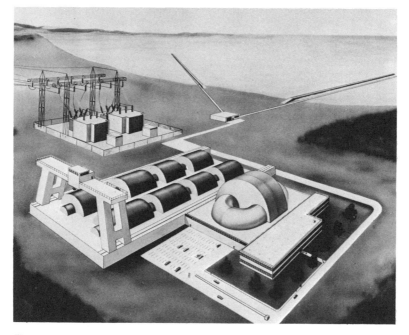

Fig. 11. A fusion power generating plant of the future. Princeton Plasma Physics Laboratory.

The Heat Problem

No one has succeeded in building a fusion reactor yet. This is because tremendous temperatures, probably between 45 and 100 million degrees centigrade, are necessary before fusion will start. This creates two big problems: first, it is difficult to heat the fuel to this temperature except in a bomb and, second, any material vessel used to contain the fuel would vaporize long before it got this hot.

Probably both these problems will be solved eventually. Scientists have been able to accelerate electrons to high speeds and pass them through deuterium or helium gas. The electrons heat the gas by colliding with molecules or atoms of the gas. At the high temperatures produced, the gas molecules break up into atoms, and the atoms separate into positively charged nuclei and negatively charged electrons. Such a mixture of equal numbers of positive and negative charges is called a *plasma*.

Magnetic Bottles

The problem of what sort of container to hold the hot plasma in is being solved by using *magnetic fields* as containers. A doughnut-shaped region in which the magnetic field runs around inside the doughnut will confine a hot plasma because the field exerts a force on the moving charges of the plasma. Such plasma-confining fields are called *magnetic bottles*.

QUESTIONS

Test your knowledge of fission, fusion, and reactors by completing the following statements. You can check your answers with those at the end of the book to see how well you understand this chapter.

1. The process in which the nucleus of an atom splits into nuclei of two different atoms is _____.
2. The process in which nuclei atoms join to form one heavier atom is _____.
3. A fission chain reaction that is continuing but is under control is called _____.
4. A chain reaction continuing and going out of control is_____.
5. A chain reaction that produces an average of one fission-producing neutron per fission is called _____.
6. A chain reaction that produces an average of less than one fission-producing neutron per fission is _____.
7. The four essential parts of a reactor are (a) _____, _____, _____, and _____ _____.
   A fifth part sometimes used to make the reactor more efficient is the (b) _____.

(TURN TO NEXT PAGE.)

8. Nuclear reactors produce _____ _____, _____, and _____ _____.
9. Fusion reactors will be better than fission reactors in three ways: (a) _____ will be cheaper and more plentiful; (b) they will be easier to _____; (c) they produce much more _____.

V. Radioactivity and Radiation

*Where do light, heat, and radio waves come from? Some kinds
of radiation are similar to light and heat; others are streams
of particles. What are some of the characteristics of these
different kinds of radiation? In some radioisotopes electrons
shoot out from the nucleus. How can an electron come from
a nucleus when there are only neutrons and protons there?
Can a person who has been exposed to radiation become
radioactive or make other people radioactive?*

Heat, light and radio waves are all forms of radiation and
many other forms, not so familiar, are being produced all the
time. Radiation has always been with us. The sun and stars
are intensely radioactive and even the earth is slightly so.
Very small amounts of naturally radioactive materials are
widely distributed in water, air, soil, plants, and animals.

The three kinds of radiation you will need to know about
are *alpha, beta,* and *gamma* radiation. Alpha, beta, and
gamma are the first three letters of the Greek alphabet.
The symbols for them are α, β, and γ. Alpha and beta radiation
consists of particles shooting out of the nuclei of unstable
atoms. Alpha particles lose energy rapidly as they pass through
air and so travel only a few inches in air. Beta particles do not
lose energy so rapidly and can travel about 25 feet.

Gamma radiation is different in that it doesn't consist of
material particles. It is a form of electromagnetic radiation,

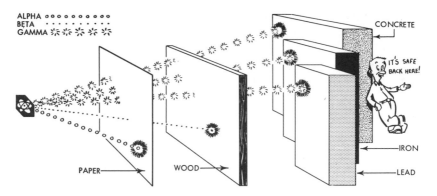

Fig. 1. Alpha radiation is stopped by a few sheets of paper; beta radiation is more penetrating. Some gamma radiation can only be reduced to safe levels by thick shields of concrete or other materials.

which means it is similar to heat, light, and radio waves. Of course, gamma radiation differs from these waves in that it can't be felt, as heat waves can, or seen, as light waves can, or heard, as radio waves can. Gamma rays can pass through much greater thicknesses of material than either alpha or beta particles of the same energy. See Fig. 1.

Radiation is the result of energy being thrown off from an unstable atom or nucleus as it regains stability. This may be easier to understand if we compare the atom in its unstable state to a rock balanced on top of a hill. We say the rock has potential energy because, if it is allowed to roll down the hill, energy will be released. When it reaches the bottom of the hill it has lost its potential energy and is in a stable state. A similar thing happens when an atom or nucleus changes from an unstable to a stable state. The potential energy it had in the unstable state is released in the form of radiation.

At this point the story becomes a little more involved since this happens in different ways for different kinds of radiation. Let's look now at some of the individual characteristics of these different kinds of radiation since understanding them is essential to intelligent safety procedures.

ALPHA EMITTING RADIOISOTOPES IN THE FORM OF DUST CAN BE IN THE AIR OR ON WORK SURFACES

ALPHA EMITTING DUST CAN ENTER BODY IN FOOD, AIR, OR THROUGH SKIN CUTS

Fig. 2. Alpha radiation.

MATERIAL PARTICLE RADIATION

Alpha Radiation

Sometimes an unstable nucleus can achieve a stable state by losing an alpha particle, which is made of two neutrons and two protons. As a comparison, let's consider a helium atom. Helium is represented by the symbol $^4_2$He. This means that most of the atoms of helium have mass number 4 and all have atomic number 2. The 2 shows that a helium atom has two protons and the 4 means the total number of protons and neutrons is four, so the nucleus must consist of two neutrons and two protons. You can see from this that an alpha particle is actually the *nucleus of a helium atom*. Since each proton has a positive electric charge and each neutron has no charge, the total charge of an alpha particle is $+2$.

An Example of Alpha Radiation. When an atom emits an alpha particle it changes to an atom of a different element because the atomic number is reduced by two. The mass number is reduced by four. An example of this is shown below.

$$^{218}_{84}\text{Po} \longrightarrow {}^4_2\text{He} + {}^{214}_{82}\text{Pb} \text{ or: } {}^{218}_{84}\text{Po} \longrightarrow \alpha + {}^{214}_{82}\text{Pb}$$

The symbol, $^{218}_{84}$Po represents the radioactive isotope of polonium with 134 (218 $-$84) neutrons in each nucleus, $^4_2$He

Compared to alpha particles, beta particles are light and fast and have more penetrating power for the same energy.

is a helium nucleus (alpha particle), and $^{214}_{82}$Pb is the isotope of lead with 132 (214 −82) neutrons in each nucleus. When this polonium isotope loses an alpha particle it changes to lead because the atomic number becomes 82.

Range and Speed of Alpha Radiation. Alpha particles are the heaviest and slowest moving of the radiation particles. They usually travel only a few inches in air and can't penetrate the skin, but they can be dangerous in spite of this. Radioactive dust or other material that emits alpha particles can enter the body through cuts or it can be in food or in the air we breathe. See Fig. 2. Alpha particles can penetrate the soft tissues inside the body. They sometimes cause a lot of damage because, having a +2 charge and moving slowly, alphas can pull many electrons away from the tissue atoms.

Beta Radiation

Beta radiation also consists of particles. They are *electrons,* sometimes called *beta particles,* that are emitted by unstable nuclei of atoms. Since beta particles are electrons they have a negative electric charge and are about 1/1840 times as heavy as a hydrogen atom. This means a beta particle is about 1/7400 times as heavy as an alpha particle.

Range and Penetrating Power. Beta particles having a certain energy move much faster than alpha particles having this same energy and they have greater range and penetrating power. They travel about 25 feet through the air but can be stopped by a sheet of aluminum ⅛ inch thick. Since they can penetrate the skin they are dangerous whether the radioactive material producing them is inside or outside the body.

Where do the Electrons Come From? Now if you are really thinking, you have some questions. Some of our statements seem to contradict each other. First we said that the nucleus of an atom is made of neutrons and protons held together by binding energy; now we say that beta radiation is electrons coming from atom nuclei. How can electrons come from a nucleus where there are no electrons?

This question puzzled scientists for a long time—so much so that some of them thought the whole theory must be wrong. It was finally straightened out when they realized that both neutrons and protons can disintegrate. They discovered a new particle, called a *positron,* which has the same positive charge as a proton and the same small mass as an electron. Sometimes a proton breaks up into a positron and a neutron. The positron only exists until it hits an electron; then both disappear in a flash of energy, the whole thing happening in a fraction of a second.

And a neutron can break up too. When a neutron disintegrates it becomes a proton and an electron and this is what actually happens in beta radiation. One of the neutrons of an unstable nucleus breaks up into a proton and an electron. The proton stays in the nucleus so the atomic number of the atom is raised by one, while the mass number stays the same. The electron is shot out of the nucleus as a beta ray. See Fig. 3.

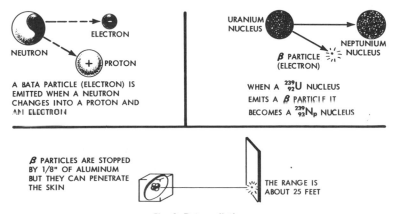

NEUTRON   ELECTRON   PROTON

A BATA PARTICLE (ELECTRON) IS EMITTED WHEN A NEUTRON CHANGES INTO A PROTON AND AN ELECTRON

URANIUM NUCLEUS   NEPTUNIUM NUCLEUS β PARTICLE (ELECTRON)

WHEN A $^{239}_{92}$U NUCLEUS EMITS A β PARTICLE IT BECOMES A $^{239}_{93}$Np NUCLEUS

β PARTICLES ARE STOPPED BY 1/8" OF ALUMINUM BUT THEY CAN PENETRATE THE SKIN

THE RANGE IS ABOUT 25 FEET

Fig. 3. Beta radiation.

Both these changes are reversible; a proton and an electron can join and become a neutron, and a neutron and a positron can combine to become a proton. This may seem strange but changes like these are always occurring in nature. Scientists have discovered more than thirty different kinds of particles, including neutrons, electrons, protons, and positrons, and they all change into other particles, gain and lose mass, or give off or absorb energy under the right conditions.

Changes in Radioactive Elements

When an atom emits an alpha particle its atomic number is reduced by two; if a beta particle is emitted the atomic number is increased by one. This is the way radioactivity causes an element to change into something else. Many atoms will absorb neutrons if they are left in a reactor. If the new isotopes formed in this way are unstable they start losing particles and change into entirely different elements.

Uranium Changes to Plutonium. As an example of how beta radiation changes a nucleus we can describe the conversion of uranium–238 ($^{238}_{92}U$) to plutonium, the man-made reactor fuel. We mentioned briefly in the last chapter that this is the process that goes on in breeder reactors. The reaction is written like this:

$$^{238}_{92}U + n \longrightarrow {}^{239}_{92}U \longrightarrow {}^{239}_{93}Np + \beta \longrightarrow {}^{239}_{94}Pu + \beta$$

Let's go over this formula and see what it means. The letter n represents a neutron. In the first part of the formula a neutron in a reactor strikes the nucleus of a uranium–238 atom and is absorbed by it. The result is another isotope of uranium: uranium–239.

Uranium–239 is radioactive. It emits a beta particle, which means one of the neutrons in the nucleus changes to a proton and a beta particle. The atom that is left has one more proton and one less neutron. Its atomic number is 93 so it is an atom of neptunium with mass number 239.

This isotope of neptunium is radioactive too. The neptunium–239 atom emits another beta particle. This reduces the number of neutrons by one again and raises the atomic number to 94, which is the atomic number of plutonium.

Disintegration of Uranium–238. A radioactive element can go through a long series of changes before it finally reaches a stable state. Uranium–238 is a good example of this. In the course of nuclear disintegration, a uranium–238 atom changes to isotopes of thorium, protactinium, radium, radon, polonium, and lead. All the isotopes are radioactive except the last one, lead 206 ($^{206}_{82}Pb$).

Other Particle Radiation

Certain atoms in an unstable state can emit neutrons, positrons, or other particles, but normally none of these represents a radiation hazard except neutron radiation. Neutron radiation occurs near nuclear reactors, neutron sources, and high-energy particle accelerators. It is as dangerous as alpha and beta radiation and much more penetrating. This is why reactors are heavily shielded to protect employees.

While an atom is disintegrating by emitting alpha or beta radiation it may also be losing energy in the form of gamma radiation. But this doesn't change the atomic number or the mass number since gamma radiation doesn't consist of particles. We will explain how this happens in the next section.

ELECTROMAGNETIC RADIATION

When Roentgen discovered X-rays he gave them that name because he didn't know what they were and X is commonly used as a symbol for an unknown quantity. Later it was discovered that they were electromagnetic waves similar to light, heat, and radio waves. Gamma rays are also a form of electromagnetic waves.

It might be easier for you to understand electromagnetic radiation if we explain where it comes from. Some atoms, or nuclei, that are in an unstable state change suddenly to a more stable state without changing their atomic weights or numbers, and in the process give off a unit of energy called a *photon*. This is very much like the case we mentioned in the introduction where a rock is balanced on top of a hill. When

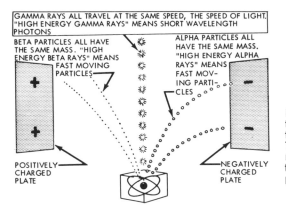

GAMMA RAYS ALL TRAVEL AT THE SAME SPEED, THE SPEED OF LIGHT, "HIGH ENERGY GAMMA RAYS" MEANS SHORT WAVELENGTH PHOTONS

BETA PARTICLES ALL HAVE THE SAME MASS. "HIGH ENERGY BETA RAYS" MEANS FAST MOVING PARTICLES

ALPHA PARTICLES ALL HAVE THE SAME MASS. "HIGH ENERGY ALPHA RAYS" MEANS FAST MOVING PARTICLES

POSITIVELY CHARGED PLATE

NEGATIVELY CHARGED PLATE

Fig. 4. The energy of particle radiation varies with the speed of the particles. The energy of electromagnetic radiation varies with the wavelengths of the photons.

it changes from its unstable position on top of the hill to the stable position at the bottom, the rock loses energy.

Photons and Material Particles

Photons act like particles in some situations and in others they act like waves, something like the waves a pebble makes when it is dropped in water. But this is not what makes them different from the other particles we have talked about. Electrons, protons, neutrons, and positrons all have this same property in reverse; that is, they act like waves in some situations.

The difference between photons and "material" particles that we can understand is related to their masses and their speeds. Electrons, protons, and all the other material particles travel at various speeds but they never go as fast as the speed of light. They have mass, or weight, at all speeds and at rest. Photons travel at the speed of light, 186,000 miles per second, and they have no mass. See Fig. 4.

The photons created when an unstable nucleus or atom goes to a more stable state travel at the speed of light until they disappear. This can happen in several ways. Sometimes they are absorbed by other atoms or nuclei, in which case the atoms or nuclei become less stable than they were before. Another way a photon can disappear is by being absorbed by a material particle. In this case its energy goes into making

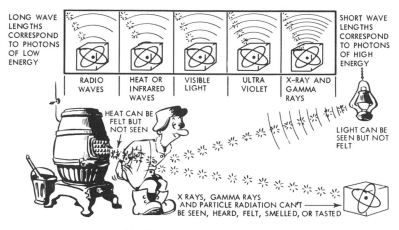

Fig. 5. Electromagnetic radiation.

the material particle go faster. These are only two of many ways a photon can change to something else.

Photons and Radiation

Light, heat rays and gamma rays are all composed of photons. These photons all move at the same speed but they are different from each other in other ways. When they are acting like particles, some carry more energy than others. When they are acting as waves, they show different frequencies and wave lengths. The photons carrying the highest energies have the shortest wave lengths.

As Fig. 5 shows, the electromagnetic waves with longest wave lengths are radio waves; heat waves and visible light have shorter wave lengths. X-rays and gamma rays have the shortest wave lengths.

Gamma Rays

Atoms will emit photons when the outer electrons change their orbits or speeds so as to put the atom into a more stable state. Often an electron moves from an orbit farther from the nucleus to one that is closer, putting the atom in a more stable state. The atom then releases its excess energy in the form of a photon. The photon, of course, may be a light, heat, ultraviolet, infrared, radio, or X-ray photon, depending on which electron was involved, its orbital position, etc.

When it is the nucleus, not the atom as a whole, that is unstable, gamma ray photons or sometimes X-ray photons, are emitted. An example will show how this works.

When a radium atom disintegrates it sends out an alpha particle which moves at one of two possible speeds. What's left is a nucleus of the heavy gas, radon. If the alpha particle is moving at the faster speed, all the excess energy of the radium nucleus has gone into propelling the particles. But if the alpha particle moves at the slower speed, the radon nucleus is left with some excess energy which makes it unstable. It can regain stability by releasing the excess energy in the form of a gamma ray photon.

Range and Penetrating Power

Gamma rays have the highest penetrating power of all forms of radiation having the same energy. They are so penetrating that most nuclear reactors need eight feet of concrete shielding to stop enough of the gamma radiation to make the outside safe. Table I summarizes the properties of alpha, beta, and gamma radiation.

TABLE I. PROPERTIES OF ALPHA, BETA, GAMMA RADIATION			
Symbol	α	β	γ
What they are	Particles. Nuclei of helium atom	Electrons emitted when neutrons in nuclei change to protons	Electromagnet waves or photons
Weight (Mass)	Heavy. Four times hydrogen atom	Light. 1/1840 as heavy as hydrogen atom	None
Electrical charge	Positive (+ 2)	Negative (−1)	No charge
Penetrating ability	Stopped by several sheets of paper	Stopped by 1/8" thick aluminum	Half of gammas stopped by 1/2 inch of lead
Speed	2,000 to 20,000 miles per second	46,000 to 170,000 miles per second	186,000 miles/sec. Speed of light
Type of hazard	High internal hazard	Moderate internal or external hazard	High external hazard

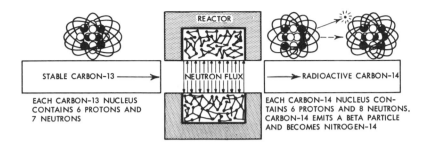

Fig. 6. If a rod of stable carbon–13 passes through a reactor, a few of the nuclei absorb stray neutrons and become radioactive carbon–14.

FISSION AND RADIOACTIVITY

Fission and radioactivity are both the results of unstable
nuclei changing to stable states. The difference is, in fission
the nucleus splits into two new nuclei while in radioactivity it
achieves greater stability by losing a particle and sometimes
one or more gamma ray photons at the same time. Fission
also releases neutrons and great quantities of heat, light, and
gamma rays.

How Radioisotopes Are Made

The nucleus of an atom may absorb one of the neutrons
fission releases and become unstable, as Fig. 6 shows. This is
why many materials become radioactive if they are left in nuclear reactors. More radioisotopes are now being produced in
this way than by any other method.

There are two other ways of producing useful radioisotopes. When uranium and plutonium atoms split, the resulting lighter isotopes, called *fission products,* are radioactive.
Some of these can be used and the others are disposed of.
Materials can also be made radioactive if they are bombarded
with neutrons or other particles in particle accelerators.

Can People Become Radioactive?

Now that you understand what fission and radioactivity

are, you can see how difficult it would be for a person who has received a dose of radiation to become radioactive or make other people radioactive. Of course, if a person took large enough amounts of radioactive material into his body by breathing or swallowing, the radiation emitted might have enough penetrating power to reach another person in sufficient quantities to be dangerous.

A person whose clothing was contaminated with a radioactive substance could contaminate another person by contact. But this is not because the person is radioactive himself. Theoretically, of course, you could become radioactive if the atoms in your body absorbed neutrons and became radioactive, for example, by being close to an exploding nuclear bomb or some other neutron source.

HALF LIFE

Half life is a term used to show how long a particular radioactive material will last. It means the length of time in which half of the radioactive material will change to another isotope. For instance, radioactive thorium has a half life of 8×10^4 years, which means, if you had one pound of this material, half of it would decay to other elements in this time, leaving a half pound of thorium. See Fig. 7.

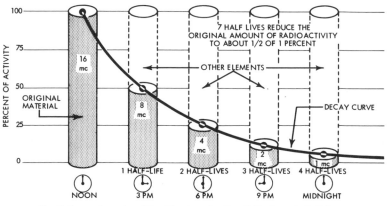

Fig. 7. Curve shows how 16 millicuries of radioactivity decay by half-lives.

At the other end of the scale, ^{199}Au, one of the radioactive isotopes of gold, has a half life of 3.3 days. If you have a pound of this gold you had better get rid of it fast. In 3.3 days there will be only a half pound left and in another 3.3 days there will be only a quarter of a pound.

QUESTIONS

Try to answer the following questions to check your understanding of radioactivity. Then check your answers with those at the end of the book.

1. The three types of radiation important to you are _____, _____, and _____ radiation.
2. The radiation that consists of nuclei of helium atoms is _____ radiation.
3. The radiation that consists of electrons is _____ radiation.
4. The radiation that is an electromagnetic wave, like heat and light, is _____ radiation.
5. _____ radiation is the least penetrating.
6. The fastest and most penetrating radiation is _____ radiation.
7. Beta radiation occurs when a (a) _____ in an unstable nucleus breaks up into a (b) _____ and an (c) _____.
8. When an alpha particle leaves a nucleus the atomic (a) _____ of the atom is reduced by 2 and the mass (b)_____ is reduced by (c) _____.
9. When a beta particle leaves a nucleus, the atomic _____ of the atom is increased by 1.
10. In the following reaction, fill in (a) the atomic number and (b) the mass number of the plutonium (Pu) isotope formed when a neptunium atom emits a beta particle.
$$^{239}_{93}\text{N}_\text{P} \rightarrow (a)\_\_\_\_\_\text{Pu}^{(b)}\_\_\_\_\_ + \beta$$
11. Do the same thing for the isotope of oxygen (O) produced by alpha radiation in Rutherford's famous transmutation reaction:
$$\alpha + {}^{14}_{7}\text{N} \rightarrow (a)\_\_\_\_\_\text{O}^{(b)}\_\_\_\_\_ + \beta$$
12. Iron–55 ($^{55}_{26}$Fe) is a radioisotope with a half life of four years. If you had a pound of it, it would take (a)_____ years to reduce this to $\frac{1}{4}$ pound and (b)_____ years to reduce it to $\frac{1}{8}$ pound.

VI. Measuring Radiation

Radioactivity is measured in curies; radiation is measured in roentgens, rads, and rems. Why are so many units necessary? How can handling a radioisotope with a three foot rod make the difference between a safe dose and a dangerous overdose of radiation? If you want to cut your radiation exposure exactly in half should you cut your time of exposure in half or move twice as far from the source as you were before? What are the easiest ways to protect yourself from radiation?

Most of us could not give the exact, scientific definition of an inch, but we know that it is a unit of length and that twelve of them make a foot, and we can picture in our minds about how long an inch is. We need to know the same kinds of things about the units for measuring radioactivity and radiation. The exact definitions of the terms—curie, roentgen, rad, and rem—wouldn't mean much to us but we should have a practical working knowledge of them.

Before defining the radioactivity and radiation units we will go over some of the more common terms of the metric system of measurement, since these terms will be needed later in connection with the units. The metric system is actually less complicated than the English system we usually use. This is because there is no simple relationship among the units of the English system—a pound is 16 ounces, a yard is 3 feet, a foot is 12 inches and in general, larger units are multiples of various numbers of smaller units. In the metric system each smaller unit is multiplied by 10 or a power of 10 to get the larger units. This makes calculation simpler because to multi-

TABLE I. THE METRIC SYSTEM		
METER(S)	METRIC UNIT	EQUIVALENT INCHES
1000	Kilometer	39.37×10^3
1/10	Decimeter	3.937
1/100	Centimeter	.3937
1/1000	Millimeter	.03937
1×10^{-6}	Micron	3.937×10^{-5}

TABLE II. STANDARD METRIC PREFIXES	
PREFIX	MEANING
Deci-	1/10
Centi-	1/100
Milli-	1/1000
Micro-*	1/1,000,000
Kilo-	1000
*The term micron is used instead of "micrometer".	

ply a whole number by 10 you just add a zero at the end of it, and to multiply by a higher power of 10 you add two or more zeros.

THE PREFIXES OF THE METRIC SYSTEM

A meter, the basic unit of length in the metric system, is equal to 39.37 inches; a centimeter (cm) is 1/100 meter or .3937 inches; and a millimeter (mm) is 1/1000 meter or .03937 inches. In all these units of length, *meter* is the basic unit and the prefixes, *centi-* and *milli-*, are used to show what fraction of a meter is meant. See Table I.

The standard prefixes with their meanings are listed in Table II. You are probably familiar with most of them already. In all scientific measurement, whether the unit is meter, gram, or some other unit, the same system of prefixes is used. In other words, if you saw the term milliroentgen you would know that it means 1/1000 roentgen, even if you didn't know what a roentgen was. In the same way, you would know that a microcurie is 1/1,000,000 curie. The metric prefixes are even used with English units in some cases. For example, in very precise shop work we sometimes speak of a *micro-inch* which is, of course, 1/1,000,000 inch.

QUESTIONS

To check your understanding of the metric prefixes, fill in the following blanks or write the answers on a separate sheet, and then check your answers with those at the end of the book.

1. Since a meter is 39.37 inches, a kilometer is _____ inches.
2. A decimeter is (a) _____ times as long as a centimeter and
   (b) _____ times as long as a millimeter.
3. A milligram is _____ gram.
4. Thirty mm is about _____ inches.
5. Ten cm is _____ inches.

CURIES, ROENTGENS, REMS, RADS

You have four new terms to learn in this chapter—*curie, roentgen, rem,* and *rad.* These are basic units used to measure radioactivity and radiation. Sometimes it is necessary to know how many atoms in a radioactive substance will disintegrate in a certain time. In a case like this we would use the curie unit. At other times it might be more convenient to know something else about the radioisotope, such as the total amount of energy lost by the isotope's radiation as it passes through air or how much damage this would do to a human being. Other units are needed for these cases. This is why there are several different units of measurement. We will start with curies.

Curies

One curie is defined as that quantity of any radioactive material in which 3.7×10^{10} atoms are disintegrating every second. This number is used because the curie unit is based on the radioactivity of radium. In one gram of radium 3.7×10^{10} atoms disintegrate every second, so one gram of radium contains one curie of radioactivity.

A curie of some other material could be less than a gram or it could be much more. For example, as Fig. 1 shows, one curie of phosphorus–32 ($^{32}_{15}P$) weighs 20 grams. The amount depends only on how fast the unstable atoms give off particles or gamma ray photons.

A Measure of Rate. Curies can be confusing when you first hear about them if you are used to units of measurement that do not depend on the materials being measured. For example, an inch is always the same length whether you are measuring wood, metal, or any other material. But the same number of curies might be contained in different amounts of different materials. If you are driving at the rate of 60 miles per hour it doesn't matter whether you are on a motor scooter or in a heavy Cadillac, the rate will be the same. The curie is a rate measurement, but it *does* depend on the materials.

Millicuries and Microcuries. In tracer work four nuclear

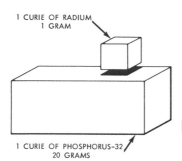

1 CURIE OF RADIUM
1 GRAM

1 CURIE OF PHOSPHORUS-32
20 GRAMS

Fig. 1. The number of grams containing a curie can vary with the radioisotope.

RADIOISOTOPE

The roentgen is a measure of radiation, not of radioactive material.

disintegrations per second can easily be counted. Compare this with the 3.7×10^{10} disintegrations per second of a curie of any material and you can see what a large unit the curie is. For this reason the smaller units, *microcurie* and *millicurie,* are usually used instead. As you know from reading the last section, a millicurie is $\frac{1}{1000}$ curie and a microcurie is $\frac{1}{1,000,000}$ curie.

A curie measures radioactivity in all types of radioactive materials, although it was originally used for those emitting gamma radiation only. This unit was adopted when the relation of disintegration to emission was not well understood. Research has shown that some radioisotopes emit more than one ray per disintegration. For example, cobalt–60 emits two strong gamma rays for each atom that disintegrates while some radioisotopes emit only one. Since a measuring instrument will count gamma rays (rather than the number of disintegrating atoms) directly, a curie of cobalt–60 will register twice as much gamma radiation as a curie of radioactive material in which a nuclear disintegration produces only one gamma ray of the same energy.

Roentgens

The roentgen is a unit of measurement of X-rays and gamma rays. The abbreviation for roentgen is r and it is pronounced *rent'ghen*. The roentgen measures how much X or gamma radiation is present at a certain point in terms of the amount of radiation needed to produce a certain standard amount of *ionization* in the air around that point. Ionization of the air occurs when photons knock electrons out of the atoms of oxygen, nitrogen, etc., in the air. Therefore, ionization produces electrically charged particles — negatively charged electrons, positively charged atoms or molecules, and other positive and negative particles.

One roentgen is defined as that quantity of X-ray or gamma radiation that will produce one electrostatic unit of positive (or negative) charge in a cubic centimenter of dry air.

There are three important limitations in the definition of the roentgen: (1) ionization produced by particle radiation such as neutrons and alpha particles is not taken into account, (2) ionization produced in substances besides air, such as hydrogen, pure oxygen, lead, human tissue, etc., are not accounted for, and (3) although ionization is being "produced" in a cubic centimeter of air, for high energy photons the electrons *produced* by the photons in a certain volume of air do *not stay* in that volume but fly out of it because the electrons themselves have a lot of energy. The roentgen can only measure how much ionization was produced by the radiation; it does not measure how much energy was needed to produce that ionization or whether the ionized particles remained in the given material.

Rads

Because of the limitations of the roentgen, a more general unit of measurement is necessary. When high-energy radiation passes through any material, energy is usually removed from the radiation. This energy is absorbed by the atoms and molecules of the material and can result in dissociation of the molecules into atoms, ionization of the atoms, etc. The energy absorbed per gram of material is called the *absorbed dose.* The unit of absorbed dose is the *rad,* defined as an absorption of 100 ergs of energy from the radiation per gram of material that the radiation passes through. To relate the rad and roentgen to each other, we can use the equation: 1 roentgen in air = 0.88 rad in water.

Rems

Rem is the unit of dose equivalent and the letters stand for *roentgen equivalent in man.* It is often called the RBE dose because it can be defined as the dose in rads multiplied by a number called the *RBE.* RBE (*relative biological effectiveness*) is a measure of how effective any particular radiation is in producing some radiobiological effects (skin reddening, baldness, death, etc.) *as compared with* the ability of X or gamma radiation to produce the same effect. For example, if the RBE of a certain special radiation is 2 for producing baldness, and if 400 rads of gamma rays will produce baldness, then only 200 rads of the special radiation would be needed to produce baldness. Since the RBE of gamma radiation relative to gamma radiation is 1, the RBE dose of 400 rads of gamma radiation is 1 × 400 rads = 400 rems. Since the RBE of the special radiation relative to gamma radiation is 2, the RBE dose, for 400 rads the special radiation is 2 × 400 rads = 800 rems. In other words, *rems* take into account both the radiation and the effect.

Now that you have a general idea of what curies, roentgens, rads, and rems are, let's consider two other factors that are important to us—time of exposure and distance from the source. First we will describe the effects of time and distance on radiation exposure separately, and then we will show how total radiation exposure is determined, taking both of these effects into account.

RATE AND TIME OF EXPOSURE

If we were to say that a certain faucet delivers one gallon of water what would you know about the faucet? We haven't really told you anything because we could have meant one gallon per minute or one gallon per hour. The same thing applies to a blower. If we say it delivers 1,000 cubic feet of air we would also have to specify whether this was in an hour,

Fig. 2. Water from a faucet (left) is measured in gallons per hour. Air from a blower (middle) is measured in cubic feet per minute. Radiation from a source (right) is measured in rads per hour.

a day, or a minute, because it certainly makes a difference. When we are interested in the delivery of water, air, radiation, or anything else we have to specify the time as well as the amount, if we want to have any useful information. See Fig. 2.

In other words, it doesn't mean much when we say a man has been exposed to a radiation source delivering gamma radiation unless we tell what the dose rate was and how long the man was exposed. If the dose is in rads and the time is in hours, the dose rate is in rads per hour (rad/hr); if in roentgens, the rate would be roentgens per hour (r/hr). This means, if the man stood near a radiation source where the dose rate was 20 r/hr for one hour he would receive 20 r $\times$ 0.88 $\frac{rad}{r}$ $\times$ 1 hour = 17.6 rads or rems, but if he stood there only a half hour he would receive 8.8 rems.

Let's take another case—this time with rems of alpha or beta radiation. Suppose a man works for $2\frac{1}{2}$ hours at a spot where he is receiving radiation at the rate of 4 rems/hr. How many rems would he have been exposed to? The number would be 4 $\times$ $2\frac{1}{2}$ so he would have received 10 rems of particle radiation.

DISTANCE FROM THE RADIOACTIVE SOURCE

Distance is an important factor in figuring your radiation exposure because radiation fans out from its source. If you are close to the source the rays are close together and you get a concentrated dose. As you move back the rays fan out. More of them will miss you because they are spreading farther and farther apart.

Squares and Inverses

Before we describe the actual calculations necessary for finding total exposure let's go over two simple mathematical terms you may have forgotten, *square* and *inverse*. The square of any number is that number multiplied by itself. When you were reading about scientific notation and powers of 10 you learned that 10 squared is $10 \times 10 = 100$. Two squared means $2 \times 2 = 4$, and it is written 2^2. Three squared or 3^2 is 3×3, 4^2 is 4×4, etc.

By the *inverse* of a number we mean that number which the original number has to be multiplied by to get 1. The inverse of any number is the fraction written with 1 over the line and the same number below the line. For example, the inverse of 4 is $\frac{1}{4}$ and the inverse of 4.5 is $\frac{1}{4.5}$. Finding the inverse of an ordinary fraction is even easier—you just turn it upside down. The inverse of ¾ is ⁴⁄₃, or 1⅓, because ¾ multiplied by ⁴⁄₃ is 1. This works because the fraction with 1 over the line and ¾ under the line turns out to be ⁴⁄₃, as you can figure out by ordinary arithmetic.

The Inverse Square

The *inverse* square of a number is the square of the inverse or the inverse of the square—they are the same. For example, the inverse square of 3 is $(\frac{1}{3})^2 = \frac{1}{9}$ or $\frac{1}{3^2} = \frac{1}{9}$; the inverse square of 4 is $\frac{1}{16}$, etc.

PROBLEMS

To check your understanding of inverses, squares, and inverse squares, try writing the inverse squares of the following numbers. Then check your answers with those at the end of the book.

1. (a) 2   (b) 3   (c) 6
2. (a) ⅜   (b) 2⅔

The Inverse Square Law for Radiation

Now that you understand what inverse square means, let's see how it applies to radiation. You probably noticed when you worked the problems above that the inverse square of a number gets smaller very rapidly as the number increases. It turns out that the intensity of radiation falls off at that rate, so you can see what a great effect your distance from the source will have on the amount you receive.

The inverse square law is a statement of the relation between intensity of radiation and distance from the source that applies to all radiation, heat and light as well as all the kinds produced by radioisotopes. It can be written: *The intensity of radiation is proportional to the inverse square of the distance from the source.* But you shouldn't let that statement scare you. It is just the mathematicians' way of stating something that is really simple, but hard to say.

Calculating Radiation Exposure

This is the way we calculate the effects of distance. The amount of your exposure is the number of rems you would receive at 1 foot from the source multiplied by the inverse square of your actual distance in feet. Written with symbols:

$$\text{(number of rems/hr of exposure)}$$
$$= \text{(no. of rems/hr at 1 foot)} \times \left(\frac{1}{\text{distance}} \right)^2.$$

An example will make this clearer. Suppose you are 1 foot from a radioisotope and receiving radiation at the rate of 100 rems per hour. If you move 2 feet away the exposure would not be $\frac{1}{2}$ (100) or 50 rems/hr as you might think. It would be $(\frac{1}{2})^2 \times 100$ or 25 rems/hr.

If you move three feet away from the same source you can

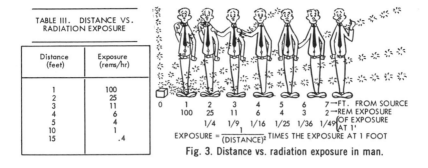

TABLE III. DISTANCE VS. RADIATION EXPOSURE	
Distance (feet)	Exposure (rems/hr)
1	100
2	25
3	11
4	6
5	4
10	1
15	.4

Fig. 3. Distance vs. radiation exposure in man.

find your exposure by multiplying 100 rems/hr by the inverse square of 3, or ⅑. Your exposure is now ⅑ × 100, or about 11 rems/hr.

Table III shows how much radiation you would receive at various distances from a source that delivers 100 rems/hr at 1 foot. This and the chart, Fig. 3, should convince you that the dangers of radioactivity drop off very quickly as the distance from the source increases. Handling a radioisotope with a 3 foot rod cuts your dose to ⅑ and moving 10 feet from the source reduces your exposure to only 1 per cent of what it would be at 1 foot. Actually, the intensity of radiation falls off even more sharply than this as you move away because some of the rays are always being absorbed by the air.

PROBLEMS

Working the following problems will help you check your understanding of the effect of distance. The answers are at the end of the book.

1. If a radiation source delivers 550 rems/hr at 1 foot what would your exposure be at a distance of 5 feet?
2. If a gamma radiation source delivers 600 r/hr at 1 foot what would your exposure be at 8 feet?
3. If your exposure to beta radiation is 10 rems/hr at 1 foot from a source, how far back must you be in order to get your dose down to 100 millirems/hr? (Remember that 1000 millirems = 1 rem, so 10 rems will be 10,000 millirems.)

When the Inverse Square Law Doesn't Work

The inverse square formula showing the fast decrease of

radiation with distance only applies when the radiation source is small and can be considered as a point, the center of the radioisotope. This is the actual situation in almost all atomic energy work. But if the radioisotope is large the rays fanning out from different points on it overlap and concentrated radiation is spread out over a large area. The same thing happens when radiation comes from a large number of scattered radioisotopes, as it does in the case of atomic bomb fallout.

DISTANCE AND TIME

We have described the effects of time and distance separately but in actual cases, of course, they are not separate problems. Let's combine these two things and solve a complete radiation problem.

A Sample Problem

If you are working 6 feet away from a radiation source that delivers 5 rems/hr at 1 foot, and you work there for $1\frac{1}{2}$ hours, what will your total dosage be? This will be easier if we write down all the facts.

The distance from the radioisotope = 6 feet.

Radiation at 1 ft. distance = 5 rems/hr (5000 millirems/hr).

Time of exposure = $1\frac{1}{2}$ hours.

The exposure per hour will be 5 rems/hr $\times$ $\frac{1}{36}$ so the total exposure over a $1\frac{1}{2}$ hour period would be $1\frac{1}{2}$ $\times$ $(5 \times \frac{1}{36})$ or about .208 rems (208 millirems).

Figuring time first—the radioisotope will deliver $1\frac{1}{2}$ $\times$ 5 = 7.5 rems to anything 1 foot away in $1\frac{1}{2}$ hours. So $\frac{1}{36}$ $\times$ 7.5 = .208 rems is the total number of rems a man 6 feet away would receive in this same time. The answer is the same, $1\frac{1}{2}$ $\times$ 5 $\times$ $\frac{1}{36}$ rems or $1\frac{1}{2}$ $\times$ 5000 $\times$ $\frac{1}{36}$ = 208 millirems whether you consider time first or distance first. A 208 millirem dose is dangerous if you get it in a short period of time.

PROBLEMS

Work the following problems to check your knowledge of the effects of time and distance on radiation. The answers are at the end of the book.

1. How many rems would you receive if you worked for 2 hours, 8 feet from a radiation source that gave off 640 millirems/hr at a distance of 1 foot?

2. How long could you work 10 feet away from a radiation source that delivers 5 rems/hr at 1 foot from the source, if you did not want to receive more than 50 millirems?

SHIELDING

In all of these cases, we have assumed that the only shielding between you and the radiation source would be the air. But in almost all actual cases workers are protected by shielding that cuts down the radiation to safe limits. Employees could be exposed to dangerous radiation only by accident. Sometimes highly trained technicians will work directly with an unshielded radioisotope but this is only done in emergency situations where a calculated risk is justified.

There are many complicated reactions among radiation particles, photons, and the atoms of the shield when different types of radiation hit the different shielding materials. The general idea is that the photons or particles collide with the electrons of the shielding atoms and are stopped or deflected. When alpha and beta particles lose their speed, they are no longer dangerous. The photons or particles that don't hit any electrons will pass right through the shield.

In most cases, an element with a high atomic number, which means a greater number of electrons, is a better shielding material than a light one. This is one reason lead is commonly used, but shields can be made of other materials also. Which material is best in a particular situation depends on many variable factors, such as the type of radiation and its intensity.

Half Value Layer

When an alpha or beta particle or a gamma ray photon

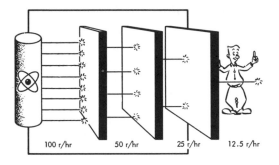

Fig. 4. If the half value layer of a material is one inch, a one inch thick shield will stop 50% of the radiation, a two inch shield 75%, a three inch shield 87.5%.

100 r/hr   50 r/hr   25 r/hr   12.5 r/hr

moves through a shield it always has a chance of getting through without hitting an electron. Shields can't be made so thick that no radiation at all can penetrate, but they are made thick enough so that the amount of radiation coming through is not dangerous to anyone.

Since shielding can't be measured according to how much will stop *all* of a particular kind of radiation, it is measured according to how much will stop half of it. Fig. 4 shows how much shielding is needed to stop 50 per cent, 75 per cent, or 87.5 per cent of any kind of radiation. The thickness of shielding that will stop half of the radiation coming at it is called the *half value layer*. It depends on what the shielding material is and also on what kind of radiation it is meant to stop.

Reactor Shields

In general, shielding thick enough to reduce high intensity gamma radiation to safe levels will stop most alpha and beta radiation completely. Neutron radiation is more penetrating than gamma rays and therefore the shields of reactors are very thick. Most nuclear reactors have shields eight feet thick which are made of a special, heavy concrete with bits of steel embedded in it.

SUMMARY

The four units of measurement are curies, roentgens, rads, and rems. A *curie* is the amount of radioactive material in which atoms are disintegrating at the same rate as they do in one gram of radium. A *roentgen* is the amount of X-ray

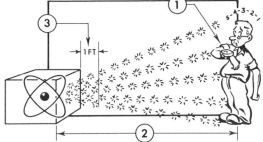

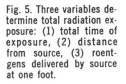

Fig. 5. Three variables determine total radiation exposure: (1) total time of exposure, (2) distance from source, (3) roentgens delivered by source at one foot.

or gamma radiation that will produce a certain amount of ionization in one cubic centimeter of dry air. A *rad* is the absorption of 100 ergs of energy per gram of absorbing material. A *rem* is the dose of rads times the effectiveness of the radiation compared with X or gamma radiation.

The prefixes used in the metric system—milli-, meaning 1/1000 and micro-, meaning 1/1,000,000—are applied to all of these units.

In figuring radiation exposure you have to consider both time of exposure and distance from the source. Suppose you know what your exposure would be if you worked at a certain distance from a radioisotope for a certain time. If you work twice as long your exposure is multiplied by 2. If you move three times as far away your exposure is multiplied by the inverse square of 3, or $\frac{1}{9}$. See Fig. 5.

QUESTIONS

To check your understanding of radiation measurement, try completing the following statements. The answers are at the back of the book.

1. The unit of radiation that is a measure of the amount of a radioisotope in which 37 billion atoms per second disintegrate is the _____
2. The unit that measures only gamma and X-ray radiation and depends on the fact that radiation produces ionization in air is the _____.
3. The unit that depends on the energy absorbed per gram of material is the _____.
4. The unit that depends on the biological damage that radiation does to the body is the _____.

(TURN TO NEXT PAGE.)

5. The inverse square of 7 is _____.

6. In practice, the *inverse square law for radiation* means: if you want to find your exposure to a source at 7 feet you would multiply your exposure at 1 foot by _____.

7. If you worked for 6 hours at a distance of 10 feet from a gamma source that gives off 200 mr/hr at 1 foot, your total exposure would be _____.

8. If you have worked for 2 hours near a radioactive source that gives off 450 millirems/hr at a distance of 1 foot, and your total exposure was 100 millirems, you were working at a distance of _____ feet.

VII. Radiation Measuring Instruments

How are geiger counters and ionization chambers read? Why do geiger counters sometimes register 0 when radiation is high? What are the instruments that can be worn by a person to measure the radiation he is receiving? How do they work?

Since radiation can't be detected by any of our senses we need instruments to tell us when it is present and how much there is. Knowing what these instruments are and how to use them is important not only for those actually working with radiation but also for those who simply work where radiation is used. See Fig. 1. Off the job, too, one might have occasion to use such instruments to measure radioactive contamination or fallout. The fallout from a nuclear bomb gives off radiation just as do radioisotopes, atomic reactors, and accelerators, but you shouldn't be afraid of atomic energy work because of this. Fallout occurs only as the result of an atomic explosion;

Fig. 1. This instrument is used for continuous monitoring of areas where radioactive materials are stored or handled. Built-in audio-visual alarms warn personnel of excessive radiation levels. Victoreen Instrument Division.

in atomic energy installations the radiation sources are confined to special areas, and safety measures prevent them from
being spread around.

MEASURING RADIATION IN AIR

The Geiger Counter

One of the most common of all radiation measuring instruments is the Geiger-Muller counter, often called the geiger
counter or G-M counter. One is shown in Fig. 2.

Basically, the geiger counter is a glass tube filled with gas
at low pressure and connected to an electric meter. When
alpha or beta particles, or gamma ray photons enter the gas
they produce an electric current that registers on the meter.

Measuring Beta Radiation. Alpha and beta particles can't
penetrate the glass of most geiger counter tubes, but many
tubes are made with special, very thin windows that can be
used to measure beta radiation. When the window is uncovered both gamma and beta rays penetrate the thin glass
and give a combined reading. To find the beta contribution,
you subtract the gamma reading of the geiger counter, measured with the window covered, from this combined reading.

How To Read a Geiger Counter. Fig. 3 shows a typical geiger
counter face. Notice that there is an ON and OFF switch and
that the ON position has three settings, $\times 1$, $\times 10$, and $\times 100$.

Fig. 2. This geiger counter is measuring
the radioactivity of allanite. The glass
tube is behind the wire screen. U.S.
Atomic Energy Commission.

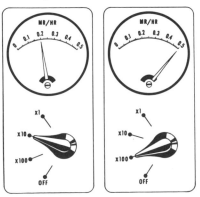

Fig. 3. Geiger counter readings.

This makes it possible to measure radiation in three different ranges. If the switch is on × 1, then the reading on the meter is multiplied by 1 to find the dose rate; i.e., the dose rate is given directly by the meter reading. For example, if the needle is on .4, your dose rate is .4 mr/hr.

If the range switch is on × 10, then the meter reading is multiplied by 10. In this case, if the meter showed .4 and the switch was on × 10, the dose rate would be .4 × 10 or 4 mr/hr. The same goes for × 100. With the switch on × 100 and the needle on .4 the dose rate would be .4 × 100 or 40 mr/hr. To check your ability to read a geiger counter, figure the dose rates shown in Fig. 3. The answers are at the end of the book.

Background Radiation. Geiger counters always register a small amount of radiation. This is not a fault of the instrument, but a true reading of the background radiation which is always present anywhere in the world. Most of the background radiation consists of cosmic rays from outer space and the rest comes from X-ray machines, naturally radioactive materials, fallout from bomb tests, and other sources.

When you turn on a geiger counter you should always get a reading between 0 and .1 mr/hr from background radiation. If the reading is 0 you should suspect the instrument isn't working properly. If it is .1 mr/hr or less you can assume the area you are testing is not radioactive.

The Blank Out Effect. In high radiation areas geiger counters will blank out from what is called *saturation*. This is because the geiger counter is built to respond only to pulses of current produced by the radiation. When the radiation is high, the pulses are produced so fast that the current is almost continuous rather than pulsating. Since the counter does not respond to a continuous radiation-produced current, the meter will read zero. Therefore, you should always be cautious about accepting a reading of zero, and especially so in an area where you suspect radiation.

To make sure you are not getting a false reading because of this effect you should start operating the geiger counter well away from the radiation area and then walk toward it, taking readings as you go. If your readings gradually get higher and then suddenly blank off you will know that you have reached a high radiation area. Then, if the geiger counter is already set at ×100 when this happens, you will need a different type of instrument, which is described next.

Ionization Chambers

Ionization chambers are similar to geiger counters, in measuring ionization produced by radiation in a gas. But ionization chambers are more accurate and their range is much greater since they are designed to measure the continuous current produced by large amounts of radiation.

Civil Defense organizations use ionization chambers that can measure up to 500 r/hr. But, as you can see from Fig. 4, the lower readings on the ionization chamber start in the range where the upper readings of the geiger counter leave off. So most ionization chambers can't register the low levels of radiation frequently found in any jobs.

The meter is read in the same way as the geiger counter meter, except that the units are r/hr instead of mr/hr. The SET position is used when the meter is being adjusted to read zero while the chamber is out of the radiation area. The ZERO knob makes the adjustment.

Like the geiger counter, the ionization chamber is pri-

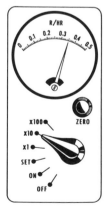

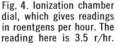

Fig. 4. Ionization chamber dial, which gives readings in roentgens per hour. The reading here is 3.5 r/hr.

Fig. 5. Scintillation counters record flashes of light caused by radiation. Tracerlab Inc.

marily used to measure gamma and beta radiation, but it can measure alpha radiation if it is equipped with a special thin (0.001 inch) alpha window in addition to the beta window. Alpha radiation is measured by taking a combined alpha, beta, gamma reading with the alpha window uncovered, then covering the alpha window and reading beta and gamma, and finally subtracting this reading from the first one.

Scintillation Counters

Another widely used counting device is the scintillation counter, an accurate laboratory instrument used by scientists. One is shown in Fig. 5. This type of counter operates on the principle that radiation striking certain chemicals causes flashes of light (scintillations) which can be converted to an electrical signal, counted, and recorded by electronic equipment. Most of us will never need to operate a scintillation counter, but we will probably see one at some time in the course of our work and we should know what they look like.

Solid State Detectors. Materials like those used in the tiny electrical circuits of transistor radios and called *semi-conductors* can also be used to detect radiation. In crystals of these materials (for example, silicon or germanium) electrons knocked free by the radiation can be collected and measured as a pulse or a continuous current of electricity. The chief advantages of solid state over gas-filled radiation detectors are their small size and high counting efficiency.

MEASURING ALPHA RADIATION

Because alpha particles have such a short range and very little penetrating power they present a special measurement problem Geiger counters, ionization chambers, and scintillation counters measure alpha radiation the same way they measure beta and gamma radiation, but the problem is to get the alpha particles into the instrument. Since they can't even penetrate a thick piece of paper, it is difficult to make a window for the counter tube thin enough to let alphas through but still strong enough to stand ordinary usage without breaking.

Solid state alpha detectors have proved very useful here, since they need no window. But even when alpha radiation can be measured, its range is so short that a low reading in one spot is no assurance that other parts of the same room are safe. To make sure an area is free from alpha radiation, you would have to cover all surfaces with the radiation counter almost touching the surface.

How Is Alpha Radiation Monitored?

The range and penetration of alpha radiation are so low that it can't do any damage as long as the source of alphas is outside the body. The main danger is from internal poisoning due to alpha radioactive dust or other materials that may enter the body in food, air, or through a break in the skin.

In laboratories or other places where radioactive alpha-emitting materials might be present, walls and other surfaces are periodically checked with thin walled counters almost in contact with the surfaces. But the radioactive material can also be in the air in the form of dust or gases. In this case it could be spread out over such a large area that the instruments wouldn't detect it.

Health physicists in radiation installations routinely check whether the air is contaminated by taking air samples with a special air pump and filter and then measuring the filter samples for alpha radioactivity with a specially constructed scintillation counter.

PERSONNEL MONITORS

Hand and Foot Counters. Fig. 6 shows a hand and foot counter that measures all three kinds of radiation. In many installations one of these machines is outside every room where an employee has any chance of getting radioactive material on his hands or feet.

Whenever anyone leaves the room he puts his hands and

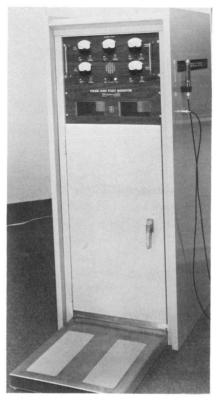

Fig. 6. A beta-gamma hand and foot monitor. Eberline Instrument Corp.

feet in the openings of the counter. This automatically starts the machine, which consists essentially of four geiger counters with walls thin enough to let in alpha particles as well as gamma and beta radiation. If there is a significant amount of radiation, a red light goes on to warn him. Then the employee goes to the health physicist's laboratory for a thorough check and, if necessary, to be decontaminated.

Although nuclear installations are constantly being checked with some version of the instruments described above, every worker in a place where there is radiation also uses some sort of personnel monitor—a small radiation detecting device worn by a person, which registers radiation whenever he is exposed to it.

Whole body counters, such as the one in Fig. 7, use two

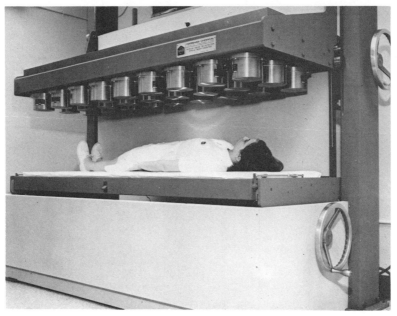

Fig. 7. Whole body counter to read whole-body exposure. Brookhaven National Laboratory.

banks of scanners, one above and one below to measure whole-body exposure.

Most personnel monitors measure only beta and gamma radiation. Since external alpha radiation is not dangerous, health physicists usually depend on hand and foot counters and on periodic checks of the areas where alpha emitters are present to control the alpha hazard. There are two basic types of personnel monitors in common use, the *film badge* and the *pocket dosimeter* (pronounced *dough sim' eh ter*).

Film Badges

The film badge is nothing more than a small piece of film wrapped to keep out light and held in a badge that can be clipped to a pocket, belt, etc. The film is sensitive to radiation and will darken in proportion to the amount of radiation it receives.

Fig. 8 shows a film badge taken apart. Notice that the film is held in a light-proof paper envelope. There is a cadmium

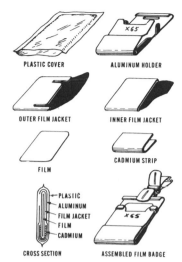

Fig. 8. Typical film badge.

strip folded over part of the film and a window through part of the aluminum metal. The cadmium strip is there so that the energy of gamma rays that hit the badge can be roughly estimated. High energy gamma rays will pass through the cadmium strip. The low energy gamma rays will not go through the cadmium and will darken only the window part of the film that is not covered by the strip.

The beta rays will only hit the portion of the film that is not covered by the aluminum. Because of this you can determine the amount of beta radiation by subtracting the total radiation registered on the film under the aluminum from the amount registered by the uncovered part.

How They are Used. If you are working regularly in a radiation area you will have a film badge assigned to you. The badge number is recorded and it is never worn by anyone else. The badges of everyone working in the radiation area are kept in a rack just outside. When you enter the area, you take the badge from the rack and put it on; when you leave you replace it in the rack. Every week the film badges are taken from the rack and the film is developed and compared to standard films that have not been exposed to radiation or have been exposed to known amounts. This comparison shows the amount of radiation that every employee has been exposed to in that week.

If the amount of radiation shown on any film is more than the amount considered safe for a week's exposure, the wearer of the badge is notified of the overexposure and given other work, where he will not be exposed to any more radiation, until the health physicist considers it safe for him to go back to the radiation area.

Pocket Dosimeters

In order to provide an immediate check on exposures, the pocket dosimeter is often worn. A person who works only occasionally in a radiation area would wear this device, since it gives an immediate indication of his radiation exposure; a film badge needs to be processed before the dose received by the worker is known. See Fig. 9.

A pocket dosimeter looks something like a pocket pen, as Fig. 10 shows, and clips in a pocket the same way. It is slightly larger than a pen and is actually a small electroscope. There are two general types, direct reading and plug-in. The direct reading type can be taken out of your pocket and read

Fig. 9. This personnel radiation monitor can be clipped on a belt or carried in a pocket. It gives an audible warning when in a gamma radiation field. Eberline Instrument Corp.

Fig. 10. A quartz fiber dosimeter. Dosimeter Corporation of America.

at any time. The other type has to be plugged into another instrument to get a reading.

The Direct Reading Type. Inside the direct reading type Fig. 11, there is a small graduated scale with a thin fiber as indicator and two fixed lenses like those of a microscope. The dosimeter is given a fixed charge of static electricity, which makes the indicator go to zero. When the dosimeter is exposed to radiation, the radiation gradually discharges it so that the

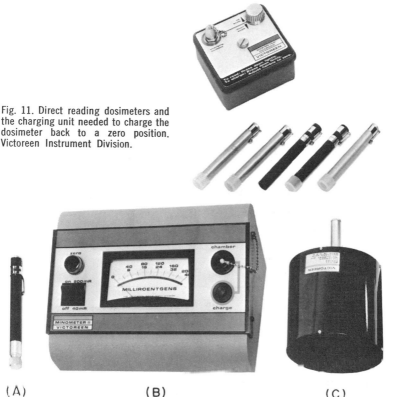

Fig. 11. Direct reading dosimeters and the charging unit needed to charge the dosimeter back to a zero position. Victoreen Instrument Division.

(A)   (B)   (C)

Fig. 12. Indirect reading dosimeters and charger-reader. (a) an indirect reading pocket dosimeter. (b) a dosimeter charger-reader. The indirect reading dosimeter is inserted in this instrument to recharge and to obtain a current reading. (c) an area dosimeter. These dosimeters are intended for the detection and measurement of low energy stray radiation. As with an indirect reading pocket dosimeter, they are inserted in the charger-reader for a reading and for recharging. Victoreen Instrument Division.

indicator starts to climb up the scale. The scale is calibrated so that the amount of radiation can be read directly from it.

Plug-in Dosimeters. In many installations where it is not necessary for employees to be able to check their exposure during the day, indirect reading dosimeters, Fig. 12, are used. They work in the same way as the direct reading type—the discharge of static electricity registers the amount of radiation—but they have to be plugged into a special charge-measuring instrument to get a reading. In one type the dosimeter itself contains only the fiber indicator. The microscope lenses and graduated scale, and a light source are in the measuring instrument. When the dosimeter is plugged in, the fiber indicator lies above the scale. The indirect reading types are less expensive and more durable than the direct reading dosimeters because their construction is much simpler.

Dosimeters are delicate instruments. Bumping or dropping them can make them discharge and register large amounts of radiation. Because of this they are often worn in pairs so that, if one is accidentally discharged, the error will be found when the two readings are compared. Just as with film badges, a worker picks up his dosimeter from the rack and wears it all day. The readings are recorded every day and a complete record of each person's total exposure is kept.

TABLE I. RADIATION MEASURING DEVICES

Type	Use	Range and Other Properties
Geiger–Muller Counters	Low level radiation counting	Under 50 mr/hr. Measure gamma and beta radiation. Will blank out in high radiation areas.
Ionization Chambers	High level radiation counting	Similar to G–M counter. Measure up to 500 r/hr. Primarily for beta and gamma radiation; some for alpha radiation.
Scintillation Counters	Accurate counting—lab use	Very precise counting device. Wide range.
Hand and Foot Counters	Check personnel contamination	Warn employees of contamination on hands, feet. Measure alpha, beta, gamma radiation.
Film Badges	Personnel Monitor	Provide permanent records of exposure. Measure alpha, beta, gamma radiation.
Pocket Dosimeters	Personnel Monitor	Provide day-to-day exposure readings and semi-permanent records. Primarily for gamma radiation.

OTHER KINDS OF RADIATION

Neutrons can't be detected by most ordinary counters but by using suitable layers of paraffin, or cadmium metal, or other materials with ordinary controls, neutron radiation can be measured. In general, special detection devices are provided whenever they are necessary in laboratories, particle accelerator installations, or reactor installations where some of the less common kinds of radiation might be present. Table I is a summary of the properties of the most common measuring instruments.

QUESTIONS

See how well you understand the material in this chapter by trying to complete the following statements. The answers are at the end of the book.

1. A geiger counter is a low level counter and usually doesn't count more than _____ _____ per hour.
2. A geiger counter may blank out on _____ level radiation.
3. An ionization chamber will count up to _____ _____ per hour.
4. The _____ counter is the most sensitive type of measuring instrument for gamma radiation.
5. Of the three common types of radiation, _____ radiation is the hardest to detect.
6. The _____ _____ provides an accurate and permanent record of personal exposure but it requires processing before reading.
7. If a geiger counter registers 0 when it is turned on it means one of two things: (a) the instrument is _____ _____, or (b) the radiation is so _____ it is out of the range of the counter.

VIII. The Effects of Radiation

Is atomic energy work dangerous? What is the safety record for employees in atomic energy installations? What are radiation injury, radiation sickness, and radiation poisoning? What are some of the visible signs of radiation injury? Does radioactivity cause human mutations? If so, what are they like? Can radiation make you impotent or sterile?

Rather than being frightened about radiation you should remember that no action is entirely safe. There is danger in everything you do. When you drive your car there is a certain chance of having an accident; after all, tens of thousands of persons are killed and hundreds of thousands are injured this way every year.

Your job involves hazards too. A construction worker is often in danger of being hit by a falling object or of falling off a building himself. An electrician has electric shock to worry about. A mechanic working around equipment with moving parts is always in danger, too. But whatever your job is, you are supposed to be aware of the hazards. You should have thought about it in advance and decided that the advantages of the job outweigh the disadvantages.

Atomic energy work is no different. When you understand the dangers involved, you can decide for yourself whether you want to do this kind of work. If you decide you can never feel safe in an atomic energy plant, you should try to find a job where your exposure to radiation will be negligible. Chances are, whatever other type of work you choose will actually be riskier than atomic energy work.

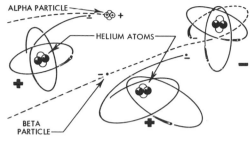

Fig. 1. An alpha particle can pull an electron out of its orbit (left), leaving a positive ion. A beta particle can knock an electron out of orbit (center), leaving a positive ion, or attach itself to an atom (right), making it a negative ion.

In the United States, the safety record of the government's atomic energy program has been better than the average of 41 other industries for 28 years (1943-1970). The overall accident rate in atomic energy during these 28 years has been consistently lower than that of industry in general. Also, the loss of time by injured workers in the atomic energy program has nearly always been lower. And in those 28 years, out of the 9,147 injuries to atomic energy workers due to *all* causes, only 38 involved radiation accidents. The rest were from the kind of accidents that happen in all construction and industrial programs.

IONIZATION EFFECTS OF RADIATION

Ionization

The ionization effects of radiation are the same for all kinds of radiation and for all materials. When radiation penetrates a material it causes some of the atoms to acquire a positive or negative electric charge. This happens because the atoms gain or lose electrons. If an atom has one electron too many orbiting around the nucleus it will have a negative electric charge. If it loses an electron for any reason, the number of protons will be one more than the number of electrons, so the atom as a whole will have a positive charge. Atoms that have more or less than the normal number of electrons in orbit around the nucleus are called negative or positive ions or ionized atoms.

Alpha and Beta Rays. Alpha and beta particles can cause

ionization in any materials they penetrate because they are electrically charged, as shown in Fig. 1. An alpha particle has a charge of $+2$. It can pull electrons from other atoms and leave the atoms with a positive charge. Beta particles repel electrons, since they are electrons themselves. A beta particle can *push* an electron from an atom as it goes by, leaving the atom with a positive charge, or attach itself to an atom, giving it a negative charge.

Gamma Radiation Effects. Gamma ray or X-ray photons produce ionization differently. Sometimes a gamma ray hits one of an atom's electrons and disappears by giving up all its energy to the atom. The energy goes into making the electron go faster so that it flies out of its orbit and leaves the atom with a positive charge. Or the photon may hit an electron and give only some of its energy to the electron, which is knocked out of the atom, while the photon continues in a new direction with lower energy. See Fig. 2.

Since the basic effects of radiation are the same on all parts of the human body for all these kinds of radiation, the only difference lies in how many atoms are ionized, what kind they are, and where they are in the body. For example, radiation can cause cataracts by ionizing atoms in the lens of the eye. The same number of ionizations would probably do no harm in a small area of the skin.

By producing electrical charges in atoms, ionization affects the chemical processes of the body. One reason for this is that some chemical compounds are broken up if the electrical character of the atoms in them changes. Since all the life processes in a plant or animal are combinations of electrical and chemical events, ionization can disrupt the normal functions of your body in many ways.

EXTERNAL AND INTERNAL SOURCES

Radiation hazards may be divided into two general categories, external and internal, depending on whether the source of the radiation is outside or inside the body. This is

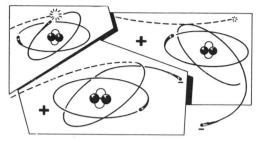

Fig. 2. When a gamma photon hits an electron (upper left), it can (1) give up all its energy to the electron. Then the electron goes faster (bottom) and flies out of its orbit. Or (2) it can give up some of its energy. Then the electron flies out of its orbit (upper right) and a lower energy photon continues in a new direction.

an important difference because low energy radiation can affect deeper tissues and organs of the body if the source is inside the body than if it has to go through the skin first. In particular, alpha emitting isotopes, which are dangerous inside the body, are no hazard at all outside it because most alpha particles do not have enough energy to get through the skin.

External radiation is easier to avoid. Its effects decrease rapidly as soon as you move away from the source. Gamma radiation is the most dangerous external source because it is a very penetrating type of radiation, but beta radiation can penetrate the skin, too. All three kinds of radiation are dangerous if they are being emitted by a radioisotope inside your body. But, since gamma and beta radiation are easy to detect you are much more likely to get an alpha emitting radioisotope into your body.

EXTERNAL RADIATION

Describing the effects of radiation is much like trying to predict the effects of falling from a high place. We can't say that if you fall 20 feet you will break your arm, but we can say you are more likely to break it than if you fell two feet.

In cases of radiation overexposure, just as with other accidents, we can't say anything about what will happen in any particular case, but we can make predictions based on the results of a large number of cases. Some people might take a dose of 250 rems with no lasting effects, while others might die from this much exposure. Still, even though some people can take much more than others, we can be *almost* certain that a person who has received a dose of 750 rems will die within a few weeks.

The Lethal Dose

In talking about the lethal effects of radiation, an abbreviation commonly used is LD_{50}. This means the *50 per cent Lethal Dose* and it is the amount of radiation that would kill 50 per cent of a large number of people subjected to it. LD_{25} means the dose that would kill about 25 per cent of the people receiving it. The LD_{50} for humans is 450 rems received over a short period of time (24 hours or less).

External Radiation Injury

If the damage a person receives from radiation is localized in one external part of the body, it is called *external radiation injury*. This type of radiation injury usually takes the form of skin burns or lesions, or loss of hair. It is most often the result of beta radiation, since the more penetrating gamma rays affect deeper regions of the body.

Skin. When this happens the skin becomes red, dry, and brittle. It is easily broken and healing from wounds is very slow. Small, wartlike growths may develop after a while. There are some indications that overdoses of radiation increase the likelihood of skin cancer, which can later spread to other parts of the body.

Hair. Loss of hair can result from exposures as low as 100 rems. It is usually confined to the areas where normal baldness occurs, such as the top of the head, but it can occur anywhere on the body. Hair loss is important only as an indication of exposure. The hair starts to grow back within a few months and no cases of permanent hair loss due to radiation exposure have been reported.

Blood Damage

The Danger of Infection. One of the immediate results of an overdose of radiation is a marked decrease in the white cell count of the blood. Since the white cells fight disease and infections, the victim of an overdose often dies from some other disease, such as bacterial infection, that his system is not strong enough to fight without the white cells. Another function of the white cells is to carry poisons out of the body. If poisonous waste products accumulate in the body they can cause illness and even death.

Blood Clotting. Radiation also causes changes in the chemistry of the blood that interfere with blood clotting. If the blood doesn't clot properly, internal and external hemorrhage become a real danger. This effect can last for several years.

Radiation Sickness

If a person receives a large enough dose of radiation there is usually a period of nausea and vomiting within a few hours afterward. Following this there is a latent period when no ill effects are apparent. The latent period may be only a few minutes for heavy doses, such as 700 rems, or it can be as long as two or three weeks for an exposure to 100 rems. After this the sickness resumes. The symptoms may include loss of appetite, nausea, loss of hair, hemorrhage, diarrhea, pallor caused by decrease in the number of red blood cells, and general emaciation.

Within the range of 200 to 600 rems, moderate to severe radiation sickness will probably occur. For doses of about 250 rems or less, the victim is likely to recover within three to six months, although this depends on the individual.

Long Term Effects

Some of these effects of radiation can last a long time after the actual sickness has passed. The condition of the skin that makes it easily broken and slow to heal, anemia and decreased resistance to infection due to blood damage, and general weakness can persist for months. In addition there are other effects that may not show up until months or years after the exposure.

Sterility. One of the most common fears of men is that

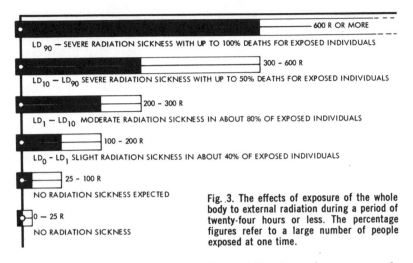

600 R OR MORE

LD $_{90}$ — SEVERE RADIATION SICKNESS WITH UP TO 100% DEATHS FOR EXPOSED INDIVIDUALS

300 - 600 R

LD $_{10}$ — LD $_{90}$ SEVERE RADIATION SICKNESS WITH UP TO 50% DEATHS FOR EXPOSED INDIVIDUALS

200 - 300 R

LD $_1$ — LD $_{10}$ MODERATE RADIATION SICKNESS IN ABOUT 80% OF EXPOSED INDIVIDUALS

100 - 200 R

LD $_0$ - LD $_1$ SLIGHT RADIATION SICKNESS IN ABOUT 40% OF EXPOSED INDIVIDUALS

25 - 100 R

NO RADIATION SICKNESS EXPECTED

0 — 25 R

NO RADIATION SICKNESS

Fig. 3. The effects of exposure of the whole body to external radiation during a period of twenty-four hours or less. The percentage figures refer to a large number of people exposed at one time.

radiation will make them sterile. While there is some truth in this, most of the common beliefs about radiation and sterility are not true. For example, many people confuse sterility with impotence. Impotence is the inability to engage in normal sexual relations, while *sterility* is the inability to have children. Radiation can cause temporary sterility but it has no effect on potency. Also, sterility has nothing to do with changes in masculine or feminine characteristics. Radiation exposure never has any effects of this kind.

Radiation can cause *permanent* sterility in both men and women, but this requires a very large dose, 500 to 625 rems, in a single, localized, short exposure. If delivered to the whole body, such a large dose would cause death in roughly 8 out of 10 cases. However, doses of 100 rems or more, localized to the reproductive organs, could cause *temporary* sterility without any other sickness.

Leukemia. One of the possible delayed effects of an overdose of radiation is cancer. Leukemia, which is blood cancer, is one of the most frequent types; and another is skin cancer.

Leukemia can result from radiation because bone marrow is very sensitive to radiation damage, and blood cells are manufactured in the bone marrow. One of the immediate results of an overdose of radiation is a decrease in the number of white blood cells. After this effect wears off there can be other disturbances in the normal function of the bone marrow. In leukemia it produces too many white cells. This leads to complications that could eventually cause death.

Fig. 3 is a list of the probable effects of different amounts of exposure. The figures in the table are based on a whole body exposure received in a short period of time, which means 24 hours or less. The 24 hour figure is chosen just to make discussion easier. If you received the same total exposure in 21 hours or in 27 hours, it probably wouldn't make much difference.

INTERNAL RADIATION

Radiation Poisoning

Radiation poisoning is any illness caused by radioactive materials inside the body. Radioisotopes enter the body the same way other poisons do, in food or air, or through cuts or abrasions in the skin. In your work you would be most likely to be accidentally poisoned by alpha emitters, since it is not difficult to keep radiation areas free from the radioisotopes that emit beta and gamma radiation. Radiation poisoning can also be caused by intense fallout from nuclear bomb testing.

How Radioisotopes Collect in Certain Organs

Some radioisotopes are difficult or impossible to remove from the body after they get in. This is because your body selects substances by their chemical properties, not their nuclear properties. It can tell the difference between most elements by their chemical properties but if atoms are the same chemically—in other words, if they are different isotopes of the same element—or if they are very similar, they are treated exactly alike.

The Thyroid Gland. For example, the thyroid gland needs iodine and takes it out of the bloodstream as the blood passes through. If some of the iodine atoms are radioactive it makes no difference. They collect in the thyroid gland with the other iodine atoms. Because of this, radioactive iodine can be useful as a tracer. Doctors can give patients a small amount and then check the thyroid for radioactivity later. They can tell how well the gland is functioning by the amount of radioactive iodine concentrated there. But if a very large amount of radioactive iodine enters the body it can damage the thyroid, and this in turn affects the functioning of the whole body.

Radium and Radioactive Calcium

In the same way, since the bones take calcium out of the bloodstream, radioactive calcium will collect in the bones. Also, radium has chemical properties so much like calcium that the bones will take it from the blood along with calcium. Then the radioactive calcium or radium becomes lodged in the bone marrow where it interferes with the production of blood cells and causes various disorders of the blood.

EFFECTS ON THE POPULATION

Life Span

It is generally accepted now that, considering a large number of people, radiation will shorten their average life span, but there is no agreement as to exactly how much. One study indicates that the reduction will be about five days for every rem of radiation the people are exposed to during their lives. The figure from another study is half this amount, 2.5 days for every rem of exposure.

Using the higher figure of five days for every roentgen, we can figure that, since the average person receives about 20 rems during his life from X-ray therapy and from cosmic rays and other sources of natural radioactivity, his life span is about three months shorter than it would be if he were never exposed to radioactivity. The average atomic worker receives about .1 rem per year. This means he accumulates an additional 3 rems during a thirty year working life, and his life span is reduced by fifteen days.

Of course, if a worker received 5 rems per year (the maximum permissible dose) for thirty years his life span would be decreased by about two years, but this much exposure is very unlikely. Also, these figures deal with the average results from a large group of people and can't be applied to any particular case. An atomic energy worker may live to be 100 or die at 25; no one can predict the life expectancy of a single individual.

Genetic Effects

Genetics is the science of heredity and the genetic effects of radiation are its effects on our children and our descendants. To understand the genetic effects of radiation we will have to know a little bit about how we inherit traits from our parents and ancestors.

Heredity. Every male or female sex cell contains 23 fibers, called chromosomes, and each chromosome is made up of thousands of smaller units called *genes*. Each gene represents a certain physical characteristic such as height, color of hair, or color of eyes. When one male and one female cell unite to produce a new individual, each chromosome from the mother pairs off with one contributed by the father and the genes representing the same trait match up. In other words, everyone has two genes determining each of his physical characteristics, one from his mother and one from his father.

Dominant and Recessive Traits. Some traits, such as black hair, are *dominant* and others, like red hair, are *recessive*. This means, if one parent contributes a gene for black hair and the other a gene for red hair, the child will have black hair. In order to get a red haired offspring both genes have to be those producing red hair. Any time an individual has two different genes for one trait, the dominant gene determines which characteristic he will have.

Since a recessive trait can't show up unless the individual has two genes for it, recessive genes can be carried by a family for generations without showing up. Dominant traits always show up but recessive traits may be overpowered for an indefinite period of time.

Mutations. Sometimes something happens to one or more of the genes and this introduces an entirely different characteristic into the offspring. Such a change is called a *mutation*. The present normal rate of mutations in human beings ranges from about 1 to 100 for every million births, depending on the type of mutation. Some mutations might be useful because they introduce new and desirable traits into the race. On the other hand, experts agree that at least 90 percent of all human mutations are harmful, since most mutations will not work in properly with other normal human characteristics.

Mutations do not produce "monsters." They are often such slight changes that they can't be detected under ordinary circumstances. For example, a mutation might take the form of increased or decreased resistance to certain diseases. A person who had this new trait would never know it unless he were exposed to the diseases. Also, most mutations are recessive. This means, if a person's genes were altered for any reason, the new traits might not show up for several generations, and his children would be unlikely to have them.

Mutations and Radiation. Although it is now generally accepted that radiation increases the rate of mutation, the amount of increase is hard to determine for human beings. One reason is that the effect of a mutation might not show up for fifty or a hundred years. Even if we could be sure that someone had a characteristic caused by mutation and that the mutation had been caused by radiation, we still couldn't know whether it had originated in one or both of his parents or in his remote ancestors. Also, interpreting statistics is always a problem.

Because of these difficulties, most estimates of the genetic effects of radiation on human beings are based on studies of animals and insects that reproduce quickly, rather than on direct evidence. These animals supply a large number of generations in a short time and they inherit characteristics in the same way we do, except that the number of chromosomes is different. By applying their results with animals to human heredity, experimenters have been able to reach some useful conclusions.

Radiation does not produce any new kinds of mutations. It only increases the number of the same ones we have always had. The amount of increase that could be caused by atomic energy work is very small. It has been estimated that to double the present rate of mutation every member of the population would have to receive an additional 40 rem dose. The average atomic energy worker should receive only about 3 rems in thirty years as a result of his work. This is much less than the exposure from one fluoroscopic examination (10 to 20 rems).

QUESTIONS

See if you understand the effects of radiation by trying to complete the following statements. Compare your answers with those given at the back of the book.

1. The three harmful effects of radiation are radiation _____, radiation _____, and radiation _____.
2. LD_{50} means _____ _____ for _____ per cent of the people exposed to a hazardous situation.
3. The LD_{50} for radiation is _____ _____.
4. Although radiation effects vary with the individual, it is almost certain that a dose of _____ rems will cause death within a few weeks.
5. When we are referring to radiation exposure, a short period of time means _____ hours or less.
6. The two types of isotope radiation that present an external hazard are _____ and _____ radiation.
7. _____ radiation is only dangerous if the isotope emitting it is inside the body.
8. In a 28-year period the accident rate has been consistently _____ in atomic energy installations than in other industry employing the same number of people.
9. The _____ properties of a radioisotope determine where it will go in a person's body.
10. The inability to have children is called (a) _____; the inability to have normal sexual relations is called (b)_____.

IX. Protection from Radiation

*Who decides how much radiation is a safe dose? What safety
measures are necessary in installations using atomic energy?
Why are fires and explosions especially dangerous when radioactive materials are present? What are the treatments for radiation poisoning both in use now and in the development stage?*

Most laboratories and industries use nuclear radiation
sources obtained from and carefully controlled by the Atomic
Energy Commission (AEC). AEC regulations are so strict
that radiation accidents and overexposures are extremely rare.
Still, nothing is foolproof. As the use of radioactive materials
becomes more widespread, the chances are that safety measures
will be relaxed. It is only common sense to regard all radiation
with caution and avoid any unnecessary exposure.

At the present time you have more to worry about from
medical radiation facilities than in most other installations
using atomic energy. Medical uses of radiation cannot be
restricted the way industrial uses are, and it is possible to
receive more radiation in one medical examination than an
atomic energy worker would receive in several years. But here
there is usually no question about the advantages of the medical examination outweighing the radiation risks. About two
thirds of all the radiation the average person in the United
States receives from all sources comes from medical examina-
tions and treatments. Table I shows about how much radioactivity you can expect to receive from medical and other
sources.

One of the most dangerous things about radiation is the

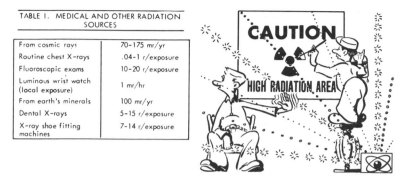

TABLE I. MEDICAL AND OTHER RADIATION SOURCES	
From cosmic rays	70–175 mr/yr
Routine chest X-rays	.04–1 r/exposure
Fluoroscopic exams	10–20 r/exposure
Luminous wrist watch (local exposure)	1 mr/hr
From earth's minerals	100 mr/yr
Dental X-rays	5–15 r/exposure
X-ray shoe fitting machines	7–14 r/exposure

Fig. 1. Because there is no pain, some people adopt an I-can-take-it attitude about radiation.

fact that there are no visible effects immediately after exposure. No one has to tell you to let go of a hot electric line or to stop banging your finger with a hammer. The pain will make you stop. But radiation doesn't cause pain; you could receive a dangerous overdose and not even feel it until much later. See Fig. 1. Only *knowledge*—knowledge of safe limits for exposure; of the use of measuring instruments, and of the effects of time, distance, and shielding—can protect you from radiation injury the same way pain protects you in other dangerous situations.

The best general safety practice anyone can follow is to understand radiation hazards himself so that he doesn't have to depend on other people to tell him when to be careful. If you have something to gain by exposing yourself to a small amount of radiation you will want to be able to decide for yourself whether it is worth the risk. In the case of atomic energy work, employees receive so little radiation exposure that the advantages of these high-paying jobs usually far outweigh any disadvantages.

SAFE LIMITS FOR EXTERNAL RADIATION EXPOSURE

The maximum permissible doses (MPD) of radiation are recommended by the Federal Radiation Council (FRC) and have become enforceable by law through the Occupational Safety and Health Act. For workers in atomic energy the MPD is about thirty times the low dose everyone receives from natural sources of radiation such as cosmic rays. MPD's are based on past experience with radiation and try to strike a balance between risk and benefit. Like any other standards, arbitrary limits are set that permit the work to be done but still insure the safety for the worker. Once these limits are determined, they are enforced rigidly.

A parallel example of a standard is the Building Code Standards. It has been determined that 2×4 studs in a wall set at 16-inch centers provides a reasonable amount of strength for a reasonable cost. It might be possible to build a strong house with studs on 18-inch centers but this would not pass inspection under the Building Code. The Code would soon lose its meaning if exceptions were allowed until gradually houses would be built with studs on 24-inch or 36-inch centers.

The Maximum Permissible Dose. The Maximum Permissible Dose (MPD) as set by the Federal Radiation Council has been written into the new Federal Occupational Safety and Health Standards Act (OSHA). Details of this act can be found in the Federal Register, Volume 37, Number 202, Section 1910.96. This act makes the following provisions:

1. Maximum ionizing radiation during any one 3-month period:

Type of Exposure	Rems Per Calendar Quarter
Whole body; Head and trunk; active blood-forming organs; lens of eyes; or gonads	$1\frac{1}{4}$
Hands and forearms; feet and ankles	$18\frac{3}{4}$
Skin of whole body	$7\frac{1}{2}$

2. The MPD for persons under 18 years of age is 10% of the amounts shown in the above item 1.
3. The MPD listed in item 1 above can be exceeded under the following conditions:

   A. The dose to the whole body does not exceed 3 Rems during any calendar quarter and the total lifetime dose that the individual has received does not exceed 5 (M—18) Rems, where "M" equals the individual's age in years at his last birthday.

   B. The employer must maintain past and current exposure records for each individual which show that the dose to the individual will not exceed the amount described above.

What A and B, *above* mean is that the whole body exposure of 1¼ Rems per quarter can be increased to 3 Rems per quarter provided accurate records are available to show that the individual will not receive a lifetime total dose to his whole body that exceeds 5 Rems for each year since his 18 birthday. For example, a person who is 24 years old may not receive a total lifetime dose of over 6 × 30 Rems. The 6 is determined by subtracting 18 from 24. Therefore, if an employer has records to show that a 24-year old employee has only been exposed to a total dose of, say, 10 Rems, that employee could be permitted, if necessary, to work in a restricted area where the whole body dose was up to 3 Rems in a calendar quarter.

The Once-In-A-Lifetime Dose

It has been suggested that a once-in-a-lifetime dose of 25 rems be allowed an individual for carrying out extremely important work, such as an emergency rescue in case of accidents. This is in addition to the 5 rems per year allowance and is a sort of "bonus" that can be used only once in a person's lifetime. A person who has received this 25-rem dose is taken off radiation work, at least temporarily.

PROTECTION FROM INTERNAL RADIATION

The seriousness of radiation poisoning depends on what material is emitting the radiation as well as how much and what kind of radiation it is. Some radioisotopes will lodge in an organ of the body and stay there indefinitely, such as radium in the bones, while others are rapidly eliminated.

Biological Half Life and Effective Half Life

As you learned before, the radioactive half life of a radioisotope is the time it takes for half the material to change into other elements. The *biological half life* of a radioactive material is the time it takes for half of the material to leave the body by natural chemical and biological processes. The *effective half life* is the time it takes for the combination of radioactive decay and natural elimination to reduce the amount of radioactive material in the body by half. The effective half life depends on both the biological half life and the radioactive half life and it is a shorter time than either one of them. Table II shows how these three kinds of half lives are related to one another.

Treatments for Radiation Poisoning

Radiation poisoning is an extremely difficult problem that is just now beginning to be attacked. As you have seen, the damage any radioisotope will do internally depends on three things: its radioactive half life, its biological half life, and its chemical properties, since they determine which organs the radioisotope will go to in the body. Since nothing can be done to make a radioactive substance lose its radioactivity or to change its chemical properties, the treatments are all methods for getting the radioisotope out of the body while at the same time treating some of the symptoms such as vomiting and diarrhea and also making up body-fluid losses so that the patient does not lose strength. In addition, the natural body processes of recovery and repair are assisted by giving the patient bed rest, good care, and good nutrition.

TABLE II. COMPARISON OF
HALF LIVES

Radioactive Half Life (years)	Biological Half Life (years)	Effective Half Life (months)
1	1	6
1	2	8
1	3	9
2	1	8
2	2	12
2	3	14
3	1	9
3	2	14
3	3	18

Treatment with Competing Atoms. One of the most successful methods of preventing damage by radiation poisoning is to introduce into the body a substance that is chemically similar and that will compete with the radioisotope. This works only if the competing element is taken into the body before the radioactive material has been deposited in the organs that use it.

For example, large quantities of calcium salts can be taken immediately after a dose of any of the bone-seeking radioisotopes, such as radium, radioactive calcium, or strontium 90. If large amounts of normal calcium are in the blood at the same time, most of the radioisotopes can be prevented from lodging in the bone marrow. This works because the bone marrow can't tell the difference and accepts some of the calcium atoms instead of the radioactive atoms. If the blood offers ten times as much normal calcium to bone as radioactive substitutes, about ten times as many normal calcium atoms as radioactive atoms will be accepted.

Fluid Treatments. One cure that you might almost think is worth the disease is the method of eliminating those isotopes that tend to lodge in fluids of the body. In order to increase the flow of water through the body and wash those isotopes out before they can lodge in permanent places, large quantities of beer or soda pop are drunk.

Chelating agents are compounds that combine with radioisotopes that are insoluble in water. The resulting compounds are soluble in water and can be carried out of the body this way. The most successful of the chelating agents has the long name, ethylenediaminetetracetic acid (called EDTA for short) .

SUMMARY OF SAFETY PROCEDURES

Now that you understand what radiation is and what it can do to your body tissues, most of the standard safety procedures will seem to you to be just common sense. At the present time there is no safer place to work than an atomic energy plant; nevertheless, your safety depends to a large extent on your ability to follow the directions of the health physicist and to use your own knowledge to avoid blundering into a dangerous situation.

Radiation Symbols

Fig. 2 shows the radiation warning sign that must be displayed in all radiation areas and special labels that are used for shipping radioactive materials. The sign has a yellow background and the propellor-like figure is colored purple. Whenever you see this sign you should exercise caution, especially in hospitals or medical facilities where large exposures are possible. No one should enter an area marked with the radiation symbol without specific instructions from his supervisor.

Radioactive materials are shipped and stored in shielding containers that reduce the radiation to a safe level. In addition, fissionable materials are provided with containers designed to keep two or more subcritical masses from coming together to form a critical or supercritical mass. These materials are

Fig. 2. Radiation warning sign (left) and shipping labels (center and right).

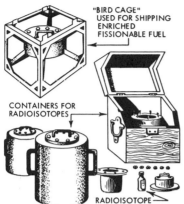

Fig. 4. A radioisotope is prepared for shipment. This container consists of a lead shield in a wooden box. Note tongs and counter. Brookhaven National Lab.

Fig. 3. How radioactive materials are shipped and stored.

always marked with special labels that include information on the kind of hazard they represent and the appropriate safety measures. See Figs. 4 and 5.

General Safety Procedures

Whenever you are working in a radiation area you should find out from the health physicist the nature and location of the radiation source, and the safety precautions you are expected to take. It is his job to see that all the work is done safely and that everyone knows how to protect himself from overexposure.

Film badges or pocket dosimeters are usually provided in radiation areas and in some places where there may be radioactive contamination in the air, respirators are also used. These should be worn at all times when you are in the area and, if a hand and foot counter is provided, you should never leave without using it.

Time, Distance, and Shielding. Remember that distance works for you and that moving only a few feet can cut down your exposure tremendously. Handling a radioisotope with a three foot rod can make the difference between a safe dose and a dangerous overdose. Time works for you too, although not as much as distance. The shorter the amount of time you spend in a radiation zone, the less exposure you will receive.

Some radioisotopes are shielded only in certain directions when they are in use. If you don't understand what the shield is for and that you are supposed to stay behind it, you can inadvertently receive an overdose by being in an unshielded position.

Radioactive Contamination. In an area where radioactive contamination is possible you should always act as if arsenic were spread over everything, and wash thoroughly as soon as you leave. Of course, alpha emitters and other radioisotopes are not as dangerous as arsenic, but they enter the body in food, through breaks in the skin, or by being inhaled in the same way as other poisons. This is the reason you are not allowed to eat, drink, or smoke in a radiation area. Vacuum cleaners are always used instead of brooms for cleaning up so that the alpha emitters and other radioisotopes will not be scattered into the air. In some radiation areas special clothing is worn. Laboratory coats and coveralls are widely used and sometimes caps, shoe covers, canvas or rubber gloves, and masks or respirators are also provided.

Part of the job of the health physicist is to see that radioactive wastes from nuclear reactors, laboratories, and other installations are disposed of safely. Liquid and solid wastes are trapped and stored in isolated areas until they can be disposed of. If the material has a short half life it can be handled safely within a relatively short period of time. Materials with long half lives have to be stored for longer times.

Accidents

Any radiation area is required by the Occupational Safety and Health Act (OSHA) to have an alarm system that goes off automatically if radiation increases beyond a safe level. When this sounds it means *get out fast*. Of course, you will want to find out before you start working how to get out of the area in the shortest possible time.

In the case of a fire or explosion it is especially important to get clear of the radiation zone even if you are not in a place where you can be injured, because fire and smoke may spread radioactive contamination over a large area. Since health physicists are trained to take care of all accidents, you should always leave emergency measures to them, or else wait outside the danger area until they can arrive and supervise rescue or other emergency operations.

QUESTIONS

1. The Occupational Safety and Health Act has set the maximum permissible whole body dose of radiation that a radiation worker can receive at _____ rems per year.
2. The MPD for exposure of hands is _____ rems per quarter.
3. In case of an emergency, a person may be permitted a "once-in-a-lifetime" dose of _____ rems.
4. The time it takes for a person's body to eliminate half of a foreign substance is called the _____ _____ _____ of the substance.
5. The best way to protect yourself from radiation or anything else that is dangerous is to _____ it.
6. To avoid radiation poisoning you should never _____, _____, or _____ in a radiation area and you should clean up with a _____ _____ instead of a broom.
7. The man to check with about proper safety precautions is the _____ _____.

8. At the present time there is more danger of overexposure around _____ radiation facilities than around laboratory or industrial radiation sources.

Answers

CHAPTER I
1. nuclear, atomic   2. split   3. (a) steam, (b) electricity
4. radioactive   5. radioisotopes   6. tracers, or tagged atoms
7. mutations   8. fuel

CHAPTER II

1. atom   2. atomic   3. weighed   4. element   5. compound
6. nuclear physics   7. X-rays   8. mass, energy

CHAPTER III

Writing Numbers with Powers of Ten
1. (a) 200,000   (b) .0005   (c) .0000067
   (d) 8,540,000   (e) 340,000,000   (f) 1,420,000
2. (a) 3.4×10^{-3} (b) 1.003×10^3 (c) 4.52×10^6 (d) 1.03×10^{-4}
3. .0000000135   4. 1×10^{16} 5. 16, 17, 18, 19

Multiplying and Dividing with Powers of Ten and Estimating Answers
1. (a) 2.88×10^{25} (b) 2.88×10^{17} (c) 2.88×10^{-25}
2. (a) 3×10^{17} (b) 3×10^{25} (c) 3×10^{-17}
3. (a) 5×10^9 (b) 5×10^1 (c) 5×10^{-4}
4. (a) 3×10^{-1} (b) 3×10^7 (c) 3×10^2
5. (a) 6.923×10^7 (b) 6.923×10^1 (c) 6.923×10^{-5}
6. (a) 7.25×10^6 (b) 7.25×10^8 (c) 7.25×10^1

General Questions
1. electrons   2. electron   3. nucleus
4. proton   5. neutrons, protons   6. protons
7. (a) neutrons, (b) number, (c) mass
8. (a) 65, (b) 30, (c) 30, (d) 35, (e) 30

CHAPTER IV

1. fission   2. fusion   3. critical
4. supercritical   5. critical   6. subcritical

7. (a) fuel, moderator, shielding, control rods, (b) reflector.
8. electric power, radioisotopes, reactor fuel.
9. (a) fuel, (b) control, (c) heat.

CHAPTER V

1. alpha, beta, gamma   2. alpha   3. beta   4. gamma   5. alpha
6. gamma   7. (a) neutron, (b) proton, (c) electron

8. (a) number, (b) number, (c) 4 9. number
10. (a) 94, (b) 239 11. (a) 5, (b) 10 12. (a) 8, (b) 12

CHAPTER VI

The Metric System

1. 39,370 or 3.937×10^4 inches 2. (a) 10, (b) 100
3. 1/1000 4. 1.18 5. 3.937

Inverse Squares

1. (a) $\frac{1}{4}$ (b) $\frac{1}{9}$ (c) $\frac{1}{36}$
2. (a) $\frac{8}{3} \times \frac{8}{3} = \frac{64}{9} = 7\frac{1}{9}$ (b) $\frac{3}{8} \times \frac{3}{8} = \frac{9}{64}$

Distance Problems

1. 22 rems/hr 2. 9.38 r/hr
3. $\left(\frac{1}{\text{distance}}\right) \times \left(\frac{1}{\text{distance}}\right) \times 10{,}000$ millirems = 100 millirems.
   The answer is 10 feet.

Time and Distance Problems

1. (2 hrs) $\times$ $(\frac{1}{8})^2$ $\times$ (640 millirems/hr) = 20 millirems
2. 5 rems = 5000 millirems
   5000 millirems $\times$ (hours) $\times$ $(\frac{1}{10})^2$ = 50 millirems.
   The answer is 1 hour.

General Questions

1. curie 2. roentgen 3. rad 4. rem or *rad equivalent in man*
5. $\frac{1}{49}$ 6. $\frac{1}{49}$ 7. 12 millirems
8. $2 \times 450 \times \left(\frac{1}{\text{distance}}\right)^2 = 100$ millirems
   The answer is 3 feet.

CHAPTER VII

Fig. 2 (left) 2 mr/hr, (right) 50 mr/hr
1. 50 mr (50 milliroentgen) 2. high 3. 500 r (500 roentgen)
4. scintillation 5. alpha 6. film badge 7. (a) not working (b) high

CHAPTER VIII

1. sickness, injury, poisoning 2. lethal dose, 50 3. 450 rems
4. 750 5. 24 6. beta, gamma 7. Alpha
8. lower 9. chemical 10. (a) sterility, (b) impotence

CHAPTER IX

1. 1.5 (5,000 millirem) 2. $18\frac{3}{4}$ 3. 25
4. biological half life 5. understand
6. eat, drink, smoke; vacuum cleaner 7. health physicist 8. medical

Index